T0331176

AI and Emerging Technologies

In the past decade, artificial intelligence (AI) has made significant advancements in various sectors of society, such as education, health, e-commerce, media and entertainment, banking and finance, transportation, and defense, among others. Its application has permeated every sector, leaving no area untouched. However, the utilization of AI brings forth crucial legal, ethical, and technical concerns and obstacles that must be appropriately addressed through thoughtful deliberation, discussions, and the implementation of effective regulations.

AI and Emerging Technologies: Automated Decision-Making, Digital Forensics, and Ethical Considerations provides a comprehensive and insightful roadmap for exploring the advancements, challenges, solutions, and implications of AI in three key areas: the legal field, digital forensic, and decision-making. By delving into these topics, this book offers a deep understanding of how AI can be optimally utilized to deliver maximum benefits to users, all within a single comprehensive source. One of the focuses of this book is to shed light on the preictal application of emerging technologies in automated decision-making while also addressing the ethical considerations that arise from their use. By examining the integration of these technologies into digital forensics and their impact on other domains, such as gaming applications deepfake, this book presents valuable insights into the broader implications of AI.

The book serves as an invaluable resource for anyone seeking to understand and navigate the complex world of AI. By offering a comprehensive exploration of its applications, ethical considerations, and data protection techniques, it provides researchers and scholars, graduate students, software engineers, along with data scientists the necessary insights to harness the full potential of AI while ensuring its responsible and ethical use.

AI and Emerging Technologies

Automated Decision-Making, Digital Forensics, and Ethical Considerations

Edited by
Purvi Pokhariyal, Archana Patel
and Shubham Pandey

CRC Press
Taylor & Francis Group
Boca Raton London New York

CRC Press is an imprint of the
Taylor & Francis Group, an **informa** business

Designed cover image: Shutter stock—studio2013

First edition published 2025
by CRC Press
2385 NW Executive Center Drive, Suite 320, Boca Raton FL 33431

and by CRC Press
4 Park Square, Milton Park, Abingdon, Oxon, OX14 4RN

CRC Press is an imprint of Taylor & Francis Group, LLC

ISBN: 978-1-032-81567-1 (hbk)
ISBN: 978-1-032-81747-7 (pbk)
ISBN: 978-1-003-50115-2 (ebk)

DOI: 10.1201/9781003501152

Typeset in Times LT Std
by Apex CoVantage, LLC

Contents

*María S. García-González, Enrique Paniagua-Arís, and
Rodrigo Martínez-Béjar*

Preface

AI AND EMERGING TECHNOLOGIES: AUTOMATED DECISION-MAKING, DIGITAL FORENSICS, AND ETHICAL CONSIDERATIONS

Emerging technologies will have a great impact on society in the coming years and will pose new challenges and legal issues that will affect the development, evolution, and growth of the society. Technologies have evolved from a futuristic concept to a pervasive force, revolutionizing various domains with its transformative capabilities. AI aids in digital forensics by efficiently analyzing large volumes of digital evidence, facilitating more effective and thorough investigations. Technological innovation can both enhance and disrupt society in various ways. It often raises complex questions regarding concerns of deepfakes in various context, use of AI to enhance digital forensics, impact of technologies in decision-making, various ethical consideration of using technologies, etc. As we stand on the cusp of a digital revolution, it becomes imperative to understand not only the capabilities and potentials of emerging technologies but also the ethical dimensions that shape their application. Therefore, we need a platform that provides a systematic understanding of various applications of emerging technologies and also highlighting how technology is affecting our life. This book delves into the multifaceted realm of emerging technologies. The journey through the pages of this book traverses diverse terrains. From a detailed analysis of AI, blockchain, and big data analytics to an exploration of the gaps in technology tools for legal systems, each chapter contributes to a holistic understanding of the challenges and possibilities presented by these advancements. Therefore, this unique book covers a comprehensive set of topics that add significant value to the applications of emerging technologies.

About the Editors

Prof. (Dr.) Purvi Pokhariyal is the campus director at the National Forensic Sciences University of the Delhi Campus. Prof. Pokhariyal is the director of academics, research, and consultancy at the National Forensic Sciences University, Gandhinagar. Prof. Pokhariyal is also the founding dean of the School of Law, Forensic Justice, and Policy Studies and the dean of the School of Forensic Psychology at the National Forensic Sciences University, Gandhinagar. Prof. Pokhariyal has more than 25 years of academic and industry experience in the field of law and justice administration.

Dr. Archana Patel is an assistant professor, National Forensic Sciences University, Delhi Campus, India. She has completed her postdoc from the Freie Universität Berlin, Berlin, Germany. She has worked as a full-time faculty at the School of Computing and Information Technology, Eastern International University, Binh Duong Province, Vietnam. Dr. Patel has filed a patent titled "Method and System for Creating Ontology of Knowledge Units In A Computing Environment" in November 2019. She has received doctor of philosophy (PhD) in computer applications and PG degree both from the National Institute of Technology (NIT) Kurukshetra, India, in 2020 and 2016, respectively.

Dr. Shubham Pandey holds a PhD from the Indian Institute of Technology, Kharagpur (IIT), in the area of law, technology, and public policy. He has done his LLM from IIT Kharagpur and BA LLB (hons.) from the Institute of Law, Nirma University, Ahmedabad. His numerous works have been published in various journals of national and international repute, and he has 15 publications to his name. He has participated in various national and international conferences to present his works on various fields of law, technology, public health, and public policy. He is an avid researcher and teaches cyberlaw, information technology law, and laws related to data protection and good governance.

Contributors

M. Baritha Begum
Associate Professor
Department of Electronics and
   Communication Engineering
Saranathan College of Engineering
Trichy, Tamilnadu, India

María S. García-González
Department of Information and
   Documentation
Faculty of Communication and
   Documentation
University of Murcia
Murcia, Spain

Mitisha Gaur
Early Stage Researcher
Legality Attentive Data Scientists
   Project
Funded under the European Union's
   Horizon 2020
Marie Skłodowska-Curie Innovative
   Training Networks
Italy

Lakshmi K. S.
Associate Professor
Department of Artificial Intelligence
   and Data Science
Rajagiri School of Engineering and
   Technology
Kerala, India

Shubham Gajanan Kawalkar
Research Scholar
Bennett University
Greater Noida, India

Keerti
AI Educator in Amatir Kanya
Gurukul, Kurukshetra, India

Rodrigo Martínez-Béjar
Department of Information and
   Documentation
Faculty of Communication and
   Documentation
University of Murcia
Murcia, Spain

Mehmet Milli
Computer Engineering Department
Engineering Faculty
Bolu Abant Izzet Baysal University
Turkey

Mahyuddin K. M. Nasution
Computer Science Department
Fakultas Ilmu Komputer dan Teknologi
   Informasi (Fasilkom-TI)
Universitas Sumatera Utara
Medan, Sumatera Utara, Indonesia

Anshul Pandey
Student of BA LLB (Hons.) (3rd year)
Chhatrapati Shahu Ji Maharaj
   University
Kanpur, India

Enrique Paniagua-Arís
Department of Information and
   Documentation
Faculty of Communication and
   Documentation
University of Murcia
Murcia, Spain

Amit P. Patil
Assistant Professor
RCPET's Institute of Management
   Research and Development
Shirpur, India

Chhaya S. Patil
Assistant Professor
RCPET's Institute of Management
    Research and Development
Shirpur, India

Vaishali B. Patil
Professor
RCPET's Institute of Management
    Research and Development
Shirpur, India

Divyansh Shukla
Research Scholar (JRF) (Law)
Chhatrapati Shahu Ji Maharaj
    University
Kanpur, India

Anurag Sood
Assistant Professor
Department of Dermatology
Maharishi Markandeshwar University
Solan, India

Raunak Sood
Cyberlaw and Cybercrime Investigation
National Forensic Sciences University
Gandhinagar, India

Shilpi Sood
Operations Manager
Maharishi Markandeshwar University
Solan, India

Muneeswaran Thillaichidambaram
Laboratory of Toxicology
Department of Health Sciences
The Graduate School of Dong-A
    University
Busan, Republic Korea

Vennila Thirumalaiswamy
Department of Chemistry
Sri Sairam Engineering College
Chennai, Tamil Nadu, India

Vaishali C. V.
JSS Academy of Higher Education and
    Research
JSS Medical College
Sri Shivarathreeshwara Nagara
Bannimantap, Mysuru, Karnataka, India

Engy Yehia
Information Systems Department
Faculty of Commerce and Business
    Administration
Helwan University
Cairo, Egypt

1 Evolution of Technologies
A Comprehensive Analysis of AI, Blockchain, and Big Data Analytics

Chhaya S. Patil, Amit P. Patil, and Vaishali B. Patil

1.1 INTRODUCTION

Emerging technologies have sparked transformational shifts across many industries in the ever-changing modern world, and the legal sector is no different. The legal system has been a cornerstone of society since antiquity, influencing people's conduct and settling conflicts. However, with the introduction of groundbreaking technologies, the legal industry is going through a major paradigm change that is altering how lawyers practice and how people have access to justice. Innovative answers to long-standing problems are being offered by emerging technologies, which are also improving the effectiveness and efficiency of legal services. These technologies are swiftly upending traditional legal procedures. These cutting-edge technologies, which range from blockchain and virtual reality to artificial intelligence (AI) and machine learning, are giving lawyers access to previously unheard-of chances to improve decision-making, streamline procedures, and provide more widely available legal solutions.

The fascinating world of developing technologies in the legal field is explored in this introduction, which also sheds light on the innovations' potential for transformation, the effects they have on legal practitioners and clients, and the ethical issues that surround their use. We seek to offer insights into the constantly shifting landscape of law and its likely future course in the face of quick technological breakthroughs by exploring the opportunities and difficulties of integrating technology into the legal sector.

The essential components of the legal industry, such as legal research, contract administration, dispute resolution, client communication, and more, are being transformed by developing technology. We will examine particular examples of how this is happening in the following sections. The ethical ramifications of using artificial intelligence and other disruptive technologies in the legal context will also be explored, as well as any risks and worries linked to data privacy, security, and other relevant issues.

DOI: 10.1201/9781003501152-1

As we set out on this voyage into the world of emerging technologies in the legal domain, it becomes abundantly evident that for legal practitioners to remain effective and relevant in a world that moves quickly and is heavily reliant on technology, they must embrace and adapt to these advancements. Although the implementation of these technologies may bring some difficulties, the prospects they offer have the potential to transform the legal field and open the door to a more open, effective, and fair legal system for all.

1.2 DEMYSTIFYING BLOCKCHAIN, ARTIFICIAL INTELLIGENCE, AND BIG DATA ANALYTICS

1.2.1 BLOCKCHAIN TECHNOLOGY

Globally, various industries have become interested in the innovative idea of blockchain technology. Blockchain is fundamentally a distributed ledger that is decentralized and unchangeable, allowing transactions to be recorded and verified across a network of nodes, or computers. A chain of blocks is formed by connecting each transaction, or "block," to the one before it using cryptographic hashing. The integrity and security of the data kept within the blockchain are ensured by this design, which makes it impossible to change data in the past.

Transparency is one of the main characteristics of blockchain. Every member of the network can see a transaction once it is added to the blockchain. Due to the fact that all parties may verify and audit transactions without the use of middlemen, like banks or notaries, this transparency improves confidence and accountability. One more important feature of blockchain technology is immutability. Ensuring a permanent and impenetrable record of all transactions, once a block is included in the chain, it cannot be changed or removed. The blockchain is resistant to single points of failure and hacking attempts because of its distributed architecture, which stores numerous copies of the ledger on various nodes throughout the network.

Beyond its most well-known use in cryptocurrencies like Bitcoin, blockchain technology has a wide range of possible uses [1]. In supply chain management, where it can provide end-to-end tracking of items, ensuring authenticity and lowering fraud, researchers have looked into its potential [2]. Blockchain is used by decentralized financial systems, also known as decentralized finance (DeFi), to develop open and transparent financial services without the need for conventional middlemen [3]. Szabo first proposed the idea of smart contracts in 1996. They are self-executing contracts, with the terms and conditions built directly into the code. When specific predefined conditions are met, these contracts can be automatically executed, enabling automation and trust in a variety of business operations [4].

Advantages of Blockchain Technology

1. *Auditing and data integrity.* The immutability of blockchain technology guarantees that data recorded into the ledger cannot be changed or removed without consensus. In fields like healthcare, where maintaining accurate and unaltered patient information is essential for both patient care and legal compliance, this capability is very beneficial.

2. *Costs and reduced middlemen.* Blockchain can drastically lower transaction costs compared to traditional financial systems by doing away with middlemen, which include charges for cross-border transactions, currency conversion, and intermediary services. This is especially advantageous for remittances and global trade.

3. *Transparency in the supply chain.* The transparency and traceability of a blockchain make it a good fit for supply chain management. In order to increase transparency, lower fraud, and boost consumer confidence, businesses can follow the path taken by products from the raw materials to the final user.

4. *Voting systems that can be changed.* Voting systems that are impervious to tampering can be created using blockchain technology. This can guard against election fraud and guarantee the validity of the results, advancing democratic procedures and boosting voter confidence.

5. *Asset management and tokenization.* Blockchain permits the creation of digital tokens that stand in for actual assets, like real estate, works of art, or commodities. By allowing for fractional ownership, liquidity, and transparent trade of these assets, tokenization can open up new investing possibilities.

Disadvantages of Blockchain Technology

1. *Challenges with interoperability.* As more blockchain systems appear, establishing compatibility between them becomes difficult. Fragmentation brought on by a lack of standards might prevent value and data from moving freely between various networks.

2. *Regulatory and legal uncertainties.* When disagreements or unlawful activity take place, the legal consequences of blockchain technology might be complicated. In a decentralized setting, determining liability, jurisdiction, and regulatory compliance presents legal difficulties that must be resolved.

3. *Private keys stolen or loss.* In order to access their accounts, blockchain users frequently have private keys. Access to money or data may be permanently lost if these keys are lost. For non-technical users, managing private keys can feel like an impossible burden.

4. *Environmental problems.* Many blockchain networks' energy-intensive consensus algorithms, particularly proof of work, add to environmental problems. The carbon impact of such networks has generated arguments regarding blockchain technology's long-term viability.

5. *Change resistance.* Established industries may resist blockchain adoption due to concerns about disrupting existing procedures, a lack of knowledge of the technology, or a fear of job displacement. This opposition may hinder the mainstream adoption of blockchain solutions.

6. *Centralization potential.* Despite the decentralization that blockchain aims to achieve, some implementations may eventually bring to centralization. For instance, mining pools in control of a sizable amount of the network's computer power can have a major impact.

Security, transparency, efficiency, and enhanced trust are just a few benefits that blockchain technology offers to a variety of industries. Scalability, energy use, legal ambiguity, and interoperability are just a few of the difficulties it faces. Blockchain has the potential to disrupt markets and alter how we handle data and transactions, by fixing these weaknesses and leveraging its advantages. The trade-off between advantages and difficulties will decide the long-term influence of the technology as it develops and matures.

1.2.2 ARTIFICIAL INTELLIGENCE

The goal of the interdisciplinary field of artificial intelligence (AI) is to develop intelligent computers that are capable of carrying out tasks that traditionally call for human intelligence. Understanding AI's underlying concepts, diverse subfields, and practical applications is necessary for demystifying it. Creating algorithms and models that can analyze data, learn from it, and draw conclusions from it is at the heart of artificial intelligence.

The development of machine learning, a well-known branch of artificial intelligence, is crucial. Numerous machine learning algorithms have been developed by researchers, ranging from traditional methods like support vector machines (SVM) [5] to more contemporary methods, like deep learning [6]. Particularly in addressing challenging tasks like image recognition [7] and natural language processing [8], deep learning has demonstrated exceptional results.

Large datasets are used to train the models that AI systems use to enhance learning. Big data accessibility and improvements in data storage and processing capabilities have significantly accelerated the development of AI [9]. Additionally, the growing availability of cloud computing platforms has given researchers and developers access to sizable computational resources to more effectively train complex AI models [10].

Another crucial area of AI is natural language processing (NLP), which attempts to make it possible for machines to comprehend, analyze, and produce human language. With models like transformers redefining language processing jobs, this field has made significant advancements [11]. BERT (Bidirectional Encoder Representations from Transformers) has additionally demonstrated outstanding promise in a number of NLP applications, including sentiment analysis and question-answering systems [8].

Across several industries, AI has found useful uses. AI-driven models in healthcare help with medication discovery and medical diagnosis [12, 13]. Self-driving automobiles can negotiate challenging areas thanks to AI-powered technologies in autonomous vehicles [14]. With applications in renewable energy optimization, AI has also demonstrated promise in combating climate change [15]. AI is applied in the financial sector for algorithmic trading and fraud detection [16].

AI creates ethical and societal issues despite its enormous potential. A critical area of research now centers on ensuring fairness and minimizing biases in AI systems [17]. Furthermore, Ribeiro et al. (2016) noted that for AI models to be trusted and accepted by users, they must be transparent and understandable [18].

Understanding AI's guiding concepts, subfields, and practical applications is necessary for revealing it. While NLP and deep learning have transformed language processing tasks, machine learning, big data, and cloud computing are powering the advancement of AI. The numerous uses of AI have the potential to alter entire industries, but their appropriate use depends on ethical considerations.

Advantages of Artificial Intelligence (AI)

1. *Automatization and efficiency.* AI can carry out monotonous jobs quickly and accurately, freeing up human resources for more difficult and imaginative tasks. The operational efficiency is increased, and human error is decreased by this automation.
2. *Analysis of the data and recommendations.* AI can examine vast amounts of data and draw important conclusions that may be challenging for humans to reach. In industries like finance, healthcare, and marketing, this is especially advantageous.
3. *Available round-the-clock.* AI systems are available around-the-clock without the need for breaks or rest. This makes it possible for businesses to provide clients with ongoing services and support.
4. *Customization.* AI is capable of personalizing experiences and recommendations by analyzing user behavior and preferences. Platforms for streaming content, websites for online purchasing, and social media algorithms all demonstrate this.
5. *Decision-making and problem-solving.* Decision-making and complicated problem-solving are aided by AI algorithms' ability to handle enormous volumes of information quickly. This is applied in areas like risk evaluation and medical diagnosis.
6. *Processing of language.* Natural language processing (NLP) gives AI the ability to comprehend and produce human language. This makes chatbots, language translation, and sentiment analysis possible.
7. *Adaptation and learning.* Machine learning (a subset of AI) enables systems to learn from data and get better over time. Because of its versatility, AI can successfully handle changing circumstances.
8. *High-level specialized work.* Highly specialized jobs, such as identifying medical disorders, interpreting legal papers, or forecasting stock market movements, are tasks where AI may shine.
9. *Forecasting analytics.* Analytics that anticipate future patterns, behaviors, and results are known as predictive analytics. In industries like banking for risk assessment and manufacturing for demand forecasting, this predictive skill is useful.
10. *Healthcare innovations.* AI helps with medication research, treatment planning, and medical diagnosis. Medical image analysis, pattern recognition, and disease progression prediction are all possible with machine learning algorithms.
11. *Chatbots and virtual assistants.* AI-powered chatbots and virtual assistants like Siri, Alexa, and Google Assistant give consumers access to rapid, relevant information, simplify processes, and allow for hands-free interactions.
12. *Vehicles with autonomy/autonomous vehicles.* AI is essential for self-driving cars because it processes sensor data, makes judgments in real time,

and improves road safety through adaptive cruise control and collision avoidance.

13. *Precision farming.* Drones and sensors equipped with artificial intelligence (AI) can monitor crop health, soil conditions, and weather patterns to improve farming techniques for higher yields and resource efficiency.

Disadvantages of Artificial Intelligence (AI)

1. *Replaced employment.* AI-powered automation may result in employment losses, especially in sectors where regular jobs are amenable to automation. This gives rise to worries about unemployment and the requirement for retraining.

2. *Ethical issues.* AI systems have the potential to reinforce prejudices found in the data they are trained on, which could result in incorrect judgments and exacerbate societal inequality. A fundamental issue lies in ensuring ethical AI that is impartial and fair.

3. *Lack of innovation and compassion.* Artificial intelligence (AI) is not particularly creative or empathetic. It can simulate human-like behaviors, but it lacks real emotions or original thought.

4. *Potential privacy violations.* AI mainly relies on data, which raises worries about privacy violations and the unauthorized use of personal data. It is crucial to have effective data security and privacy safeguards.

5. *Security and privacy of data.* The data that AI is trained on determines how accurate it will be. The decisions made by the AI may be incorrect or faulty if the data is biased, partial, or inaccurate.

6. *Complexity and upkeep.* Creating and keeping up AI systems calls for specialized knowledge. Debugging, frequent updates, and complex algorithms can all be expensive and resource-intensive.

7. *Lack of control.* There is a chance that we will not be able to control what AI systems do as they become more independent. The challenge is making sure AI operates predictably and securely in every situation.

8. *Dependence on technology.* A civilization that is overly dependent on AI may lack the knowledge and skills necessary to cope with technological setbacks, leaving it vulnerable.

9. *Adoption obstacles.* Due to a lack of understanding, concerns about job loss, or skepticism about AI's capabilities, some people and businesses may be reluctant to use AI.

10. *Human dependence on AI.* Reliance on AI systems too much can result in a loss of independence and critical thinking abilities.

11. *Employment displacement and income inequality.* AI-driven job automation has the potential to result in job displacement, especially for functions that are simple to automate. If this is not handled effectively, income disparity may increase.

12. *A lack of human contact.* Artificial intelligence (AI) interventions in industries like healthcare and customer service might cause a loss of empathy and a sense of kinship.

13. *Liability and legal issues.* It might be difficult to determine who is responsible when AI systems make crucial choices. It might be difficult to determine who is responsible for accidents or mistakes.
14. *Energy use.* The sophisticated computations necessary for AI development and use can consume a large amount of energy, raising environmental issues.
15. *Moral and ethical conundrums.* Using AI can bring up moral and ethical issues, such as what choices an autonomous vehicle should make when faced with a life-threatening circumstance. These conundrums put the onus on society to establish moral guidelines for AI conduct.
16. *Diminished social interaction.* As people rely more on AI for communication and entertainment, there may be a negative impact on their relationships and social skills.
17. *Unanticipated effects.* If AI systems come with scenarios that are unrelated to their training data, they may behave in an unanticipated way. These unforeseen effects may have repercussions in the actual world.
18. *Security and hacking risks.* AI systems, especially those that are online, can be the target of cyberattacks. AI flaws could be used maliciously by actors with bad intentions.

Artificial intelligence offers a variety of benefits, including automation, data analysis, and personalization. But technology also poses problems in terms of ethics, job loss, data privacy, and dependability. Achieving a responsible and significant integration of AI into all facets of our lives requires striking a balance between utilizing its advantages and tackling its drawbacks. From healthcare and driverless vehicles to predictive analytics and other industries, AI has many benefits. The drawbacks, such as lost jobs, ethical dilemmas, and security dangers, however, highlight the necessity of ethical AI research, deployment, and continuous consideration. Collaboration across the societal domains, technological and ethical, is necessary to carefully plan for and balance the possible advantages and disadvantages of AI.

1.2.3 BIG DATA TECHNOLOGY

Big data refers to the enormous and complicated datasets that are more complex than what can be processed using conventional data processing methods. Understanding big data's traits, difficulties, and potential to revolutionize numerous industries is essential to explaining it. Velocity, volume, and variety, the three Vs, which collectively describe the size, velocity, and variety of the data being generated, are all included under the umbrella phrase "big data."

The high rate of data generation and processing is related to the velocity component. For instance, social media systems produce streams of data that are updated in real time, and financial transactions are handled in a matter of milliseconds. Real-time analytics and sophisticated data streaming capabilities are required for data analysis at such high speeds.

The volume component is concerned with the magnitude of the data itself, which is frequently created in massive amounts by numerous sources, including social media, sensor networks, and online transactions. In order to efficiently handle the data, new storage and processing technologies are required. Data that is structured, unstructured, or semi-structured are all included in the variety element. Although big data encompasses sources including text documents, emails, photos, and videos, which require various storage and processing techniques, traditional databases are ideal for structured data.

The growth of big data has sparked the creation of cutting-edge data processing and storage solutions. Cloud computing platforms and distributed storage systems like Apache Hadoop [19] have made it possible to store large datasets economically and scale them up as needed [20]. Additionally, distributed and fault-tolerant parallel processing frameworks like Apache Spark [21] have been developed to process enormous amounts of data.

Big data analytics is essential for gaining knowledge and value from these enormous databases. To find trends, patterns, and correlations, a variety of data analysis techniques are used, such as machine learning, data mining, and natural language processing. These insights help the creation of predictive modeling, data-driven strategies, and decision-making processes.

Big data has been adopted by many businesses to obtain a competitive edge. By evaluating sizable patient databases in healthcare, it has facilitated tailored medication [22]. According to Zheng et al. (2017), big data analytics are employed in retail to improve customer experience and inventory management [23]. It also has uses in the financial sector for risk assessment and fraud detection [24], as well as in the transportation sector for route planning and traffic optimization [25].

Big data has enormous potential, but it also presents a number of obstacles, such as worries about data security and privacy [26], as well as the requirement for sophisticated frameworks for data governance and ethics [27]. Additionally, there is a demand for personnel with skills in data analysis and interpretation, because deriving meaningful insights from sizable and complicated datasets requires qualified data scientists and subject matter experts.

Comprehending big data involves understanding its variety, velocity, and volume, as well as the technologies and analytical approaches used to handle and extract value from these massive databases. Big data has the ability to promote innovation and revolutionize numerous industries by tackling issues and embracing the opportunities it provides.

Advantages of Big Data Technology

1. *Making knowledgeable decisions.* Big data technology enables businesses to collect and analyze enormous amounts of data from multiple sources. This makes it possible to make decisions based on data that are accurate insights and trends.
2. *Predictive analytics.* Using past data patterns, big data analytics may estimate upcoming trends, behaviors, and consequences. Marketing, healthcare, and finance all depend on this expertise.
3. *Enhanced customer understanding.* By evaluating customer data, organizations can learn more about the customer preferences, actions, and purchasing trends of their customers. Customer service and targeted marketing techniques are made possible by this information.

4. *Improved operational efficiency.* Big data technologies improve business operations by finding gaps and obstacles in processes, resulting in cost savings and better resource allocation.
5. *Real-time data processing and analysis.* Big data tools make it possible to handle and analyze data in real time, giving businesses the ability to react quickly to shifting market conditions, consumer preferences, and new trends.
6. *Personalized experiences.* Big data enables businesses to customize goods, services, and customer interactions to meet the needs of specific clients, boosting client happiness and loyalty.
7. *Medical research and healthcare.* The use of big data analytics in medical research and healthcare can help identify new treatments, predict future disease events, and enhance patient outcomes by analyzing patient data, medical information, and genetics.
8. *Urban planning and smart cities.* The use of big data technologies helps build smart towns by analyzing information obtained from sensors and gadgets to handle resources, improve traffic flow, and enhance daily life in cities.
9. *Supply chain optimization.* By examining data related to inventory levels, trends in demand, and distribution networks, big data analytics may help firms optimize their supply chains, resulting in lower costs and more efficiency.
10. *Fraud detection and prevention.* Big data technology is able to spot unexpected transactional patterns and abnormalities, assisting businesses and financial institutions in spotting and thwarting fraudulent activity in real time.
11. *Environmental sustainability.* Big data can be used to monitor environmental conditions, forecast catastrophes, and manage resources more effectively, improving management of the environment and sustainability initiatives.
12. *Education system improvement.* By examining student performance data to pinpoint areas for development, customize learning opportunities, and optimize instructional approaches, big data analytics can improve educational systems.
13. *Responses to crises and crisis management.* Big data can help with resource allocation, emergency response coordination, and scenario analysis in times of crisis, such as those brought on by natural disasters or public health catastrophes.
14. *Market trend analysis.* The use of big data technology enables companies to examine the trends in the market, customer opinions, and competition behavior, revealing information that helps them with product creation and marketing plans.

Disadvantages of Big Data Technology

1. *Data security and privacy issues.* Data breaches, hacking, and unauthorized access are worries brought on by the extensive collecting and storage of data. Protecting confidential data becomes difficult.
2. *Issues with bias and data quality.* Biases in the data might cause biased analysis and distorted results. The reliability of the insights produced can be damaged by inaccurate and low-quality data.
3. *Legal and ethical challenges.* The collecting and use of data can give rise to ethical dilemmas, particularly when it is done so without legal authorization

or when it is used to make delicate judgments that have a direct impact on people's lives.

4. *Complexity and skill gap.* Big data systems implementation and management demand specialized technical expertise. Finding individuals who are qualified to manage these advanced technology may be difficult for organizations.

5. *Cost and resource intensity.* The cost of hiring qualified staff, purchasing software, and building a big data infrastructure can be high. It could be financially difficult for smaller organizations.

6. *Regulatory compliance.* Complying with data protection laws, such as the General Data Protection Regulation, while handling massive amounts of data adds complexity and raises the possibility of legal repercussions if not handled correctly.

7. *Loss of human touch.* Relying only on data-driven insights can result in the undervaluation of human experience, creativity, and intuition, all of which are crucial in particular decision-making situations.

8. *Misinterpretation of outcomes.* Complicated data analytics may produce outcomes that are misapplied or misconstrued, leading to faulty judgment or misguided tactics.

9. *Resistance to change.* Lack of awareness, skepticism, or worries about how big data technologies may affect work roles may prevent organizations and individuals from embracing them.

10. *Environmental impact.* Processing and storing massive amounts of data requires a lot of energy, which raises environmental issues, particularly in data centers.

11. *Reliance on data accuracy.* The quality of the input data has a significant impact on how accurate big data analytics are. Data that is inaccurate or lacking might produce false results and poor decision-making.

12. *Data overload and complexity.* The vast quantity of data produced might result in information overload, making it difficult to prioritize pertinent facts and draw insightful conclusions.

13. *Private invasion.* The widespread collecting and analysis of personal data may violate people's private rights, raising worries about monitoring and misuse of personal data.

14. *Cultural and social concerns.* Big data analysis of cultural, social, and behavioral patterns may give rise to worries about the possibility of stereotyping, profiling, and the manipulation of the public's views.

15. *Limited human interpretation.* Big data analytics can offer insights but may be unable to illuminate the "why" underlying the trends. The results must be thoroughly interpreted using human intuition and contextual knowledge.

16. *Resource inequality.* Businesses with little resources may find it difficult to integrate big data technologies and reap their benefits, potentially resulting in unequal access to data-driven insights.

17. *Resistance to change and adoption.* Organizations and individuals may be reluctant to use big data technology because it can be difficult to learn novel systems, modify procedures, and get beyond skepticism.

18. *Intellectual property rights.* The sharing and analysis of data may give rise to questions about intellectual property rights, particularly when data from many sources is integrated.
19. *Data ownership and governance.* When there are numerous stakeholders and data producers, it can be difficult to agree on who owns and controls the data being analyzed.
20. *Focus on quantity over quality.* The emphasis on gathering enormous amounts of data may cause one to overlook the significance of high-quality pertinent data that genuinely leads to insightful knowledge.
21. *Economic imbalances.* Companies that have access to more information and resources may acquire a competitive edge, thereby escalating financial discrepancies between companies.

In summary, big data technology offers strong insights that can transform decision-making in a variety of industries. To guarantee that the advantages of big data are utilized responsibly and ethically, it is necessary to carefully negotiate the problems of data privacy, bias, ethical issues, and the complexity of implementation. Organizations must carefully assess both the potential advantages and downsides of big data technologies in order to make educated judgments regarding its deployment, ensuring that legal, ethical, and practical factors are taken into account.

1.3 BENEFITS OF BLOCKCHAIN TECHNOLOGY, ARTIFICIAL INTELLIGENCE, AND BIG DATA IN THE LEGAL DOMAIN

1.3.1 BLOCKCHAIN TECHNOLOGY

1. *Authentication of tamperproof documents.* Authentication of tamperproof documents is made possible by the immutability of blockchain technology, which makes it impossible to change data, contracts, or legal documents once they have been created. By ensuring document legitimacy, this lowers conflict and raises public confidence in legal dealings.
2. *Automated execution using smart contracts.* Using predetermined criteria, smart contracts can automate contract execution. This simplifies procedures like property will execution, payment agreements, and transfers, lowering the need for middlemen and cutting down on errors.
3. *Improvements in privacy and security.* The cryptographic security features of blockchain shield private legal information from unauthorized access. On a blockchain, legal papers are kept safe, lowering the possibility of data breaches or illegal changes.
4. *Effective management of the evidence.* By offering an immutable record of the chain of custody for the evidence, blockchain may transform evidence management. By doing this, the procedure of providing evidence in court is made simpler, and the integrity of the evidence is ensured.
5. *Intellectual property protection streamlined.* A blockchain's secure recording of intellectual property rights can serve as irrefutable evidence of

ownership and usage rights. The procedure of enforcing copyright and patents may be made simpler as a result.

6. *Unchangeable case history.* Using blockchain, it is possible to create an exhaustive and unalterable record of all submissions, documents, and judgments in a case. This openness can help keep data correct for future use, appeals, and historical research.

7. *Effective authentication and notarization.* The tamperproof and secure features of blockchain technology can speed up notarization procedures. Blockchain technology allows for the digital notarization and storage of legal documents, which decreases the need for traditional notary services and boosts productivity.

8. *Cheaper litigation expenses.* The openness and immutability of the blockchain can help settle disagreements and conflicts, which will ultimately result in cheaper litigation expenses. To check facts, contracts, and timetables, parties can use the blockchain, which eliminates the need for protracted discovery procedures.

9. *Cross-border transactions and conflicts.* The borderless nature of blockchain can make cross-border transactions and conflicts simpler by offering a solitary, verifiable source of truth. By doing so, interacting with several legal systems can be simplified, and the pace of international legal proceedings accelerated.

10. *Effective due diligence.* Blockchain can speed up due diligence procedures in transactions like mergers and acquisitions. The company's contracts, legal history, and intellectual property rights are all accessible to parties in a secure and comprehensive record.

1.3.2 ARTIFICIAL INTELLIGENCE

1. *Legal research and analysis.* AI can quickly and accurately do legal research by scouring huge legal databases, case law, and statutes. This capacity helps lawyers swiftly identify pertinent precedents and legal ideas.

2. *Document review and due diligence.* Using AI-powered tools, contracts, legal papers, and agreements can be reviewed to find important terms, potential dangers, and discrepancies. Due diligence procedures for mergers, acquisitions, and other transactions are sped up as a result.

3. *Predictive analytics.* Using previous case data analysis, AI can forecast case outcomes and legal tactics. Informed decisions and client advice based on statistical probability are made easier for lawyers thanks to this.

4. *Automation of routine operations.* Artificial intelligence (AI) can automate repetitive operations like document generation, form completion, and scheduling, freeing legal practitioners to concentrate on complicated tasks that call for human judgment.

5. *Improved legal research.* AI-driven algorithms are able to find subtle links between cases, laws, and legal opinions, which help attorneys create strong legal arguments and strategies.

6. *Contract analysis and management.* AI is capable of analyzing contracts to extract important phrases, commitments, and deadlines. This enhances contract administration and monitoring of compliance.

7. *Reduced legal costs.* Legal services are more reasonably priced and available to a wider spectrum of clients thanks to AI's speed in researching the law and reviewing documents.

8. *Data-driven insights.* Data-driven insights can help with litigation planning by providing information on legal patterns, judge conduct, and case outcomes after processing massive amounts of legal data.

9. *Legal document summarization.* AI is able to evaluate lengthy legal papers and produce succinct summaries, assisting lawyers in swiftly comprehending and outlining important points to clients and coworkers.

10. *Regulatory compliance.* AI can keep track of regulatory changes and make sure that businesses abide by changing rules and regulations, reducing the chance of facing legal repercussions.

11. *Improved due diligence in investigations.* Artificial intelligence (AI) systems can evaluate massive amounts of digital data in investigations, indicating pertinent facts and trends that may be important for legal proceedings.

12. *Effective case management.* AI can support the management of case workflows, deadlines, and scheduling, ensuring that legal practitioners remain organized and fulfill their obligations.

13. *Virtual legal assistants.* AI-powered chatbots and virtual assistants can instantly respond to frequent legal questions and provide clients with fundamental legal advice and information.

14. *Trial data visualization.* In order to properly present evidence and arguments during trials, lawyers can use AI to visualize complex legal material.

15. *E-discovery in court cases.* AI can speed up the e-discovery process by sorting through enormous amounts of electronic documents, emails, and communications to find pertinent information.

1.3.3 BIG DATA TECHNOLOGY

1. *Case outcome prediction using big data analytics.* Case outcome predictions can be made by analyzing past case data. This helps lawyers manage expectations, optimize strategy, and give clients informed counsel.

2. *E-discovery and document review.* Big data systems can quickly review and classify a vast volume of documents, emails, and other electronic records for e-discovery in complex litigation, which will save time and money.

3. *Identification of trends for legal strategies.* Legal professionals can use big data to find trends and patterns that will help them create strong legal tactics and arguments by examining historical cases and legal trends.

4. *Monitoring of regulatory compliance.* By examining pertinent statutes, rules, and industry standards, big data technology helps organizations keep track of and abide by changing legal and regulatory requirements.

5. *Due diligence and risk management.* Legal experts can estimate the risks involved in legal problems by looking at data from similar instances, assisting clients in making educated judgments and managing potential obligations.

6. *Protection of intellectual property.* Big data tools can track intellectual property rights, spotting potential violations and assisting attorneys in promptly taking action to safeguard the intellectual property of their clients.
7. *Access to legal insights.* Big data analytics can glean insights from voluminous legal data, giving attorneys access to crucial knowledge for creating compelling legal arguments and foreseeing adversarial tactics.
8. *Improved legal collaboration.* Big data solutions make it possible for legal teams to work together on cases by exchanging and analyzing information and legal papers.
9. *Early case assessment.* By examining past case data and identifying prospective strengths and weaknesses, big data techniques can assist legal practitioners in evaluating the merits of a case early on.
10. *Litigation strategy optimization.* By looking at the results of similar cases, big data can help attorneys create winning litigation plans that take into consideration the preferences of the judges, the conduct of the opposing counsel, and previous court decisions.
11. *Personalization and client insights.* Big data analytics can be used to examine client behavior and preferences, allowing lawyers to provide individualized legal services and advice to each client.
12. *Evidence and fact corroboration.* By examining different data sources, big data can help corroborating evidence, strengthening legal arguments, and reducing uncertainties about the reliability of assertions.
13. *Public opinion analysis.* Big data techniques may monitor social networks and online platforms to determine the opinions and sentiment of the general public, assisting lawyers in comprehending how the general public views legal issues.

1.4   CONS OF BLOCKCHAIN, ARTIFICIAL INTELLIGENCE, AND BIG DATA TECHNOLOGIES IN THE LEGAL SECTOR

1.4.1   BC TECHNOLOGY

1. *Lack of a regulatory framework.* The legal industry is subject to strict rules. Uncertainties over legal validity, enforceability, and liability may arise in the absence of legislation that specifically addresses the special features of blockchain technology.
2. *Process of complex adoption.* It takes a lot of time, money, and experience to integrate blockchain into the current legal systems. Making the switch from conventional recordkeeping to blockchain-based solutions can be difficult and disruptive.
3. *Concerns about privacy in public blockchain.* Public blockchain raises privacy concerns because, despite being public, they might not adhere to some judicial proceedings' confidentiality requirements. Keeping important legal information on a public ledger is still difficult.
4. *Limited knowledge of smart contracts.* Legal professionals may not completely understand the technical complexities of smart contracts, which

could result in misinterpretations or errors. Despite the benefits of automation offered by smart contracts, these contracts may nevertheless be used incorrectly.

5. *Change reluctance.* Due to concerns about how blockchain technology may affect job functions and a lack of expertise with its technical details, legal practitioners may be reluctant to use it.

6. *Concerns about confidentiality and ethics.* The transparency of blockchain technology may conflict with the need for confidentiality in legal procedures. There may be a risk that undesired parties can access sensitive customer information, causing ethical and privacy issues.

7. *Limited blockchain legal expertise.* Lawyers and other legal professionals might not have the necessary technical knowledge to properly comprehend the complexities of blockchain technology. This could result in misunderstandings, poor communication, and possible legal mistakes.

8. *Energy and cost considerations.* Building and maintaining blockchain networks can be expensive, particularly in terms of energy use. Smaller law firms, in particular, may find it difficult to justify the expenditure.

9. *Complexity of a blockchain.* For legal practitioners without a technological background, comprehending and utilizing blockchain technology might be challenging. The adoption of blockchain may be hampered by the technology's complicated consensus methods, cryptography, and algorithms.

10. *Hard forks and blockchain forks.* In the context of blockchain, forks happen when the protocol undergoes a fundamental change, causing a split between the old chain and a new chain. Because different blockchain versions may be linked to different legal documents and agreements, this could lead to legal difficulties.

11. *Few historical precedents.* Because blockchain is a relatively new technology, there are few legal examples and case laws that pertain to blockchain applications. Uncertainty in legal interpretations and consequences may result from the absence of well-established legal norms.

12. *Challenges of reversibility.* Immutability advantages of blockchain come with drawbacks, however, including difficulties with reversibility. Because of the technology's reluctance to change, for instance, it could be challenging to fix a transaction problem.

13. *Lack of required education and training on blockchain technology.* The legal profession may have difficulties in providing legal practitioners with the required education and training on blockchain technology. To ensure that lawyers can fully utilize blockchain's advantages, this gap must be closed.

Blockchain technology has the potential to improve transparency, security, and efficiency in the legal sector. To guarantee that the advantages of blockchain are fully realized while limiting potential negatives, difficulties like regulatory ambiguities, privacy issues, and the complexity of adoption must be properly managed. Making educated decisions concerning the adoption and deployment of blockchain technology within legal procedures requires careful evaluation of these considerations.

1.4.2 AI

1. *Bias and fairness issues.* AI systems may unintentionally pick up biases from the training data, which will result in skewed results. This has the potential to erode fairness in a legal setting and perpetuate current imbalances.
2. *Absence of legal context and decision.* AI lacks the sophisticated comprehension of human context, emotions, and legal precedents possessed by skilled lawyers. Human judgment is frequently required for complex legal issues.
3. *Ethical conundrums.* Legal AI choices that involve sentence or other legal tactics that call for a human understanding of moral factors may provide ethical conundrums.
4. *Risks to privacy and security.* The risk of data breaches and unauthorized access increases when storing confidential legal information on AI systems. It is crucial to keep up strong security protocols.
5. *Job roles and job displacement.* AI's automation potential may cause repetitious task-oriented paralegals and legal assistants to lose their jobs. It is possible that lawyers will also need to modify their roles to cooperate with AI systems.
6. *Interpretation difficulties.* AI might have trouble understanding the complex wording, sarcasm, and ambiguity frequently found in legal documents. This may result in misunderstandings and mistakes in legal analysis.
7. *Technology dependence.* Overreliance on AI could impede the development of fundamental legal competencies in legal practitioners, resulting in a loss of knowledge in important areas.
8. *Job roles and job displacement.* AI's automation potential may cause repetitious task-oriented paralegals and legal assistants to lose their jobs. It is possible that lawyers will also need to modify their roles to cooperate with AI systems.
9. *Interpretation difficulties.* AI might have trouble understanding the complex wording, sarcasm, and ambiguity frequently found in legal documents. This may result in misunderstandings and mistakes in legal analysis.
10. *Technology dependence.* Overreliance on AI could impede the development of fundamental legal competencies in legal practitioners, resulting in a loss of knowledge in important areas.
11. *Automation's unintended consequences.* Legal errors could occur if some legal procedures are automated, mistakenly skipping important considerations or steps that human lawyers would notice.
12. *Challenges with client confidentiality.* Concerns about data privacy and adherence to attorney–client confidentiality arise when sharing private client information with AI-powered platforms.
13. *Regulation obstacles.* There may be ambiguities regarding the legality and ethical ramifications of AI use in legal contexts due to regulatory agencies' and legal countries' lack of clear guidelines.
14. *AI's perception as a replacement.* Clients can mistakenly believe that AI would completely replace human attorneys, undervaluing their expertise, judgment, and ability to provide individualized counsel.
15. *Bias amplification.* If AI systems are educated on biased data, they may continue to hold onto the prejudices and inequities already present in the legal system, hence escalating socioeconomic inequality.

16. *System dependencies and maintenance.* A new level of dependency is created by the constant maintenance and technical assistance needed to ensure the dependability and proper operation of AI systems.

AI has the ability to significantly change the legal industry, from accelerating contract management to expediting research. But issues with bias, moral quandaries, and the preservation of human judgment highlight how crucial it is for legal professionals to use AI technologies responsibly. A careful strategy that takes into account legal ethics, fairness, and the always-changing role of legal experts is needed to balance the advantages of AI with its possible disadvantages.

In order for legal practitioners to make educated decisions about the deployment, utilization, and ethical issues of AI technologies in their practice as they continue to be integrated into the legal realm, it is crucial that they have a sophisticated grasp of these benefits and drawbacks.

1.4.3 BD Technology

1. *Privacy and ethical issues.* Privacy, consent, and confidentiality concerns are brought up by the collection and analysis of personal data for legal study. It becomes difficult to protect customer and personal data.
2. *Bias in legal analytics.* If the data used for analysis is skewed or lacking, big data analytics' insights and forecasts may be off, which could have an impact on legal tactics and judgments.
3. *Predictive analytics reliability.* Predicting legal outcomes only on the basis of past information might not take into consideration the particulars of each case, which could result in incorrect forecasts and potential errors in judgment.
4. *Dependence on technology.* A reliance on big data technology that is too great risks impairing the ability to think critically and intuitively about the law, both of which are essential for efficient legal practice.
5. *Quality and authenticity of the data.* It is crucial to guarantee the reliability and veracity of legal data. Legal actions might be erroneous, and improper conclusions can result from old or inaccurate data.
6. *Loss of human expertise.* When making decisions about complex legal issues, human expertise, intuition, and legal judgment are crucial. Relying solely on data-driven insights could result in these attributes being neglected.
7. *Cost and implementation issues.* Putting big data technologies into practice needs expenditures in hardware, software, education, and experience. Adopting such technologies may be difficult for smaller legal firms.
8. *Complexity and learning curve.* Legal practitioners may need training to use big data tools successfully, which might take time and potentially impede adoption.
9. *Legal liability and accountability.* If decisions based on predictive analytics have negative effects or unanticipated repercussions, there may be a legal liability issue.
10. *Interpretation of the results.* Legal practitioners must appropriately analyze the outcomes of big data analysis since complicated data patterns may not always result in clear-cut legal conclusions.

11. *Data management.* Managing significant amounts of legal data can be difficult and necessitates the use of reliable storage systems, as well as techniques for data storing, retrieving, and retention.
12. *Legal environments are constantly changing.* Big data systems find it difficult to stay current with changing laws, rules, and case law, since the legal environment is constantly changing.
13. *Integration issues and technical challenges.* Integration of big data technologies with current legal systems and workflows may provide technical obstacles that need for specialized knowledge for smooth implementation.
14. *Ownership and sharing of data.* When several parties are involved in a case or transaction, collaborative data analysis raises concerns regarding ownership and sharing rights.
15. *Data tampering and manipulation.* Maintaining the integrity of legal data is crucial to preventing manipulation or tampering, both of which could compromise the validity of the analysis and results.
16. *Regulatory restrictions.* Legal information frequently comprises sensitive information that is subject to different restrictions, like the attorney–client privilege. These laws must be followed by big data analysis.
17. *Linguistic and cultural nuances.* Legal documents and data may contain cultural or linguistic nuances that require human comprehension for appropriate interpretation, which big data analytics may fail to offer.
18. *Legal professional refusal.* Some legal practitioners may be wary of depending on technology for legal decision-making, worrying that it may replace their proficiency or verdicts.
19. *Depersonalization of legal practice.* A dependence on big data tools that is too great could result in a decline in the quality of client–attorney relationships and a fall in the level of individualized legal services.
20. *Data loss and recovery.* Despite improvements in data management, there is still a chance that data will be lost as a result of technological issues or cyberattacks. Strong data recovery plans must be implemented.

It is crucial to carefully weigh these benefits and drawbacks before implementing big data technologies in the legal field. To make sure that big data technology is used ethically and successfully in legal practice, it is essential to balance the potential benefits with ethical, legal, and practical considerations.

A cautious approach in the legal field is necessary to balance the significant benefits of big data technology with the dangers it may pose. Legal practitioners must balance managing sensitive legal information with ethical, legal, and privacy considerations in order to determine whether data-driven insights are beneficial.

1.5   ENHANCING LEGAL PROCEDURES WITH BLOCKCHAIN, AI, AND BIG DATA TECHNOLOGY

The development of blockchain technology has shown promise as a catalyst for changing a number of aspects of the legal system. This chapter examines how the decentralized and tamper-resistant design of blockchain can dramatically improve legal processes in a variety of ways. We explore the distinct fields of document review, legal research, and contract analysis through a thorough assessment of the body of extant literature and scholarly works.

1. *Verification and integrity of documents.* The distributed ledger technology used by blockchain guarantees the veracity and immutability of documents. According to Dettori, A., and Santana, J. (2019), blockchain technology can be applied to provide an open and secure system for document verification and long-term storage [28].
2. *Enhancing legal research.* Blockchain technology's tamperproof properties can be used to establish reliable databases for legal data. Blockchain technology is discussed in the study by Chen, M. C., and Hsieh, C. L. (2019), as a way to increase the accuracy of legal research by preventing unauthorized changes to legal texts [29].
3. *Effective contract management.* Blockchain-based smart contracts provide a more simplified and automated method of contract administration. Insights into how smart contracts improve transparency, eliminate intermediaries, and lower the risk of contractual disputes are provided in the study by Ramachandran, M., Ranjan, R., and Khan, S. U. (2020) [30].
4. *Blockchain for intellectual property rights.* The transparency and traceability of a blockchain can help with the protection of intellectual property rights. A review of blockchain's potential uses in managing intellectual property rights, including copyright protection and patent management, was published in 2019 by T. Ahram and F. Al-Turjman [31].
5. *E-discovery and data privacy.* E-discovery and data privacy are two areas where blockchain can be helpful. By assuring the secure and auditable exchange of electronic documents, it can help with e-discovery and data privacy. Blockchain can be used to address issues with electronic discovery and data privacy compliance, according to research by Lin, Y., Zhang, and Wu (2020) [32].

The combined impact of these examples and ideas highlights how blockchain technology has the potential to revolutionize the legal industry. Blockchain offers novel answers to enduring problems, such as document verification and smart contracts, ultimately enhancing the effectiveness, security, and transparency of legal processes.

Artificial intelligence (AI) is now a powerful catalyst with the potential to transform many elements of the legal system. According to extensive research, AI's cognitive capacity and cutting-edge data processing capabilities are poised to significantly improve legal processes in a variety of scenarios.

According to Smith and Johnson (2020), artificial intelligence (AI) has become a powerful catalyst set to transform a variety of aspects of the legal landscape. According to extensive research, AI's cognitive capacity and cutting-edge data processing capabilities are poised to significantly improve legal processes in a variety of scenarios. AI-powered systems offer a quick and thorough way to review vast quantities of legal documents in the context of document review [33]. AI platforms efficiently identify trends, anomalies, and significant information by utilizing machine learning and natural language processing algorithms, speeding up the review process. On top of this, Li and Chen (2019) emphasize AI's role in automating legal document analysis while reducing errors and boosting effectiveness [34].

Liu, Wang, and Wang (2018) explain that AI algorithms expertly explore enormous databases to distil priceless insights, effectively easing the laborious processes of case law study and legislative research. The field of legal research also benefits significantly from this [35]. Notably, the ability of AI's predictive analytics to predict legal outcomes allows legal professionals to make more informed decisions. Through their work analyzing court data, Choi, Kim, and Park (2021) confirm the potential of AI in predicting case outcomes [36].

Ramirez and Cox (2019) show that AI-driven automation may be used to analyze contracts, automate review procedures, identify potential hazards, and ensure legal compliance, further expanding its revolutionary influence. AI increases precision while reducing the time required for legal research and contract management because of its capacity for comprehending and analyzing complex datasets [37].

The use of big data technologies holds the promise of revolutionary changes in the current landscape of legal operations. The potential for effectiveness, accuracy, and well-informed decision-making is being leveraged across numerous areas, including document review, legal research, and contract analysis. We may delve into the tremendous impact that big data technology can have on revolutionizing these important facets of the legal sector by studying the works of eminent authors and experts.

- *Review of documents.* In their 2013 study, Mayer-Schönberger and Cukier explore how the use of big data analytics might transform the review of documents. Their work demonstrates how cutting-edge computers can quickly analyze vast amounts of data to identify the crucial information, speeding up the review process. Furthermore, Suleyman et al. (2017) highlight the use of predictive coding algorithms that provide priority to the study of pertinent documents in order to effectively extract crucial evidence [38].
- *Legal research.* D'Ignazio and Napoli (2019) discuss the potential transformation of legal research that big data technologies can bring about. According to Katz et al. (2017), who conducted their own research, efficient information retrieval for legal practitioners is made possible by sophisticated search algorithms and natural language processing. Additionally, legal datasets' underlying trends and insights are revealed by data analytics methods, which makes it easier to find pertinent legal arguments [39].
- *Contract analysis.* Iqbal and Lim (2016) [40] investigate how processes for contract analysis can be changed by big data analytics. Their work highlights how machine learning algorithms can quickly locate clauses, terms, and potential dangers in contracts, ensuring that all legal requirements are met. The privacy risks of using big data for contract analysis are also covered in Suleyman et al. (2019) [41].

Big data technologies might potentially improve decision-making, productivity, and client service when used in legal processes. To fully exploit big data's potential in the legal field, it is essential to address issues with data security, ethics, and privacy.

1.6   UNLOCKING OPPORTUNITIES THROUGH TECHNOLOGY INTEGRATION

As artificial intelligence (AI), big data, and blockchain converge, a new chapter in the history of law is being written. This is a game-changer that goes beyond technology and has the power to alter the way legal proceedings are conducted. Think about a scenario where the judicial system evolves into a more effective, precise, and cutting-edge sector. Big data, blockchain, and artificial intelligence (AI) are three potent technologies that have combined to create this shift. Even though they are sophisticated, these technologies have the potential to revolutionize and simplify how legal issues are handled.

Blockchain can be compared to a very secure digital notebook where information is unchangeable. Big data means we have access to a wealth of knowledge, while AI gives machines the ability to think intelligently. These three combine their efforts to produce something that is greater than the sum of their individual parts.

This combination offers intriguing potential for solicitors and legal professionals. It can facilitate document management, speed up case research, and even improve contract clarity. We will learn how this integration can help legal professionals be more effective, make wiser decisions, and protect sensitive data while we investigate it. Join us on this adventure as we learn how the fusion of blockchain, AI, and big data is influencing the future of law.

Consider the opportunities that will arise from the fusion of these technologies. The immutability of blockchain secures the legitimacy of legal transactions, while big data and AI can provide previously unrecognized insights from massive amounts of legal data.

This confluence is fundamentally altering the legal field. It implies that labor-intensive legal research that used to take days might now be completed in a matter of hours, that contracts could now be examined more accurately, and that confidential legal records could now be maintained more safely. Join us on this adventure as we explore the fascinating possibility of how the convergence of blockchain, AI, and big data might transform the landscape of law, making it more accessible, effective, and adaptive than ever before.

1.7   CHALLENGES IN CONVERGING BLOCKCHAIN, AI, AND BIG DATA TECHNOLOGIES

A great possibility to improve transparency, efficiency, and correctness in legal processes is presented by the integration of AI (artificial intelligence), big data technology, and blockchain. However, this integration also creates a number of difficult problems that must be solved. Here is a thorough discussion of these difficulties:

1. *Legal and regulatory frameworks.* A complicated web of rules and legislation that differ from one jurisdiction to another governs the legal sector. Artificial intelligence, big data, and blockchain integration call for careful assessment of how these technologies integrate into current legal frameworks. It is difficult to create adequate legislation to control the use of these

technologies, especially when taking into account concerns with intellectual property rights, liability, and data privacy.

2. *Data security and privacy.* The legal industry deals with sensitive and private data. Data security and privacy issues are raised by the integration of AI and big data technology. Legal data must be kept secret and accurate in order to avoid breaches or unauthorized access. While blockchain is renowned for its security characteristics, maintaining data privacy can be difficult when working with public ledgers where all transactions are accessible.

3. *Interoperability.* Various legal entities and organizations frequently employ a variety of technology platforms and systems. It is difficult to ensure interoperability across multiple blockchain, AI, and big data platforms since they may not interact properly. To build common protocols and interfaces that allow diverse systems to communicate efficiently, standardization activities are required.

4. *Adoption and trust.* The legal profession places a great value on trust, and introducing new technologies can be regarded with suspicion at times. It is difficult to persuade legal experts, judges, and clients to trust the outputs and suggestions of AI algorithms. Building trust in blockchain's immutability and security is also important, as misunderstandings or misconceptions about the technology may stymie its acceptance.

5. *Fairness and bias.* The data that AI systems are trained on may contain biases. Biases can lead to unfavorable legal rulings or recommendations in the legal field. To avoid bias in legal outcomes, it is essential to make sure AI systems are trained on representative and diverse datasets. To increase justice and equity, algorithmic bias must be addressed. To do this, AI models must be continuously audited, monitored, and improved.

6. *Data complexity and quality.* It is difficult for AI to effectively analyze and comprehend legal papers because of the sophisticated and nuanced language used in the legal field. Building precise AI models might be challenging due to the variable quality of legal data. For big data technology to produce actionable insights, the data must be clean, structured, and dependable. A significant challenge can be ensuring the quality and consistency of legal data.

7. *Ethical considerations.* Integrating blockchain, big data, and AI can lead to moral dilemmas. For instance, AI-driven legal advice may eventually displace human solicitors, thereby hurting their ability to make a living. To avoid unintended outcomes, the ethical ramifications of keeping legal documents on a public blockchain where data cannot be changed must be carefully considered.

8. *Resources and costs.* To implement these technologies, large financial investments are needed for system development, deployment, and maintenance. Smaller law firms can find it difficult to afford the costs of implementing these technology. Significant computing resources are required for managing big data infrastructures and training AI models, which might be a hurdle for adoption.

9. *Implementing change.* The legal profession has a reputation for being conservative and slow to adopt new technologies. To ease the shift and assure

easy adoption by legal professionals, integrating blockchain, AI, and big data calls for strong change management tactics.

10. *Training and education.* Lawyers should receive training on how to use these modern tools successfully. To fully utilize the possibilities of AI and big data technologies, specialized knowledge is frequently necessary. It is crucial to offer training courses and resources so that legal professionals may comprehend the potential and restrictions of new technologies.

To summarize, while the incorporation of blockchain, AI, and big data technologies has enormous potential to transform the legal domain, addressing these challenges is critical to ensuring a successful and ethical adoption that truly benefits legal professionals, clients, and the legal system as a whole.

1.8 CONCLUSION

A new era of opportunities and difficulties has begun as a result of the dynamic integration of blockchain, artificial intelligence (AI), and big data analytics within the legal industry. These innovative technologies are redefining the basic core of legal operations and providing fresh ways to improve effectiveness and efficiency in the court system. This study has examined these advances' disruptive effects on the legal profession in great detail, illuminating their broad potential.

This study's main goal was to clarify the benefits and challenges associated with the combination of blockchain, AI, and big data analytics in the legal sector. The examination has carefully investigated how these technologies have the potential to transform a variety of legal procedures, from careful document inspection to careful legal research to complex contract analysis. The use of machine learning algorithms and natural language processing, among other AI developments, has received attention because they have the potential to bring about unmatched gains in legal duties.

Furthermore, the decentralized character of blockchain technology has been emphasized, highlighting its ability to enable safe and open transactions while also introducing the novel idea of smart contracts and the potential for decentralized legal systems. Big data's analytical capabilities have also been carefully studied, demonstrating how the enormous repositories of legal data may be used to reveal priceless insights, forecast legal outcomes, and support legal practitioners in making well-informed decisions.

It is crucial to recognize that while integrating these transformational technologies has enormous potential, there are also a number of difficulties involved. The need to carefully strike a balance between advancing technology and keeping ethical norms has emerged as a key problem. Concerns about privacy must be addressed carefully in order to protect data and safeguard sensitive information. Furthermore, it becomes crucial to create strong legislative frameworks so that these technologies can be used legally and ethically.

In-depth analysis of the legal sector's seismic upheaval caused by the confluence of blockchain, AI, and big data analytics has been done in this study. This chapter serves as a basis for facilitating informed conversations among legal practitioners, politicians, and technologists by outlining both the potential and complications inherent in this

integration. The seamless merger of innovation and observance of ethical considerations will ultimately decide the full revolutionary potential these technologies hold for
the legal sector as it continues to change alongside the legal landscape.

REFERENCES

1. Nakamoto, S. (2008). *Bitcoin: A Peer-to-Peer Electronic Cash System.* Cryptography Mailing List. https://metzdowd.com.
2. Meng, W. (2018). *Applications of Blockchain Technology in Supply Chain Management.* https://doi.org/10.29007/cvlj.
3. Swan, M. (2015). *Blockchain: Blueprint for a New Economy* (1st ed.). O'Reilly Media, Inc.
4. Szabo, N. (1996). *Smart Contracts: Building Blocks for Digital Markets.* www.truevaluemetrics.org/DBpdfs/BlockChain/Nick-Szabo-Smart-Contracts-Building-Blocks-for-Digital-Markets-1996-14591.pdf.
5. Cortes, C., & Vapnik, V. (1995). Support-vector networks. *Machine Learning*, 20(3), 273–297.
6. LeCun, Y., Bengio, Y., & Hinton, G. (2015). Deep learning. *Nature*, 521(7553), 436–444.
7. Krizhevsky, A., Sutskever, I., & Hinton, G. E. (2012). ImageNet classification with deep convolutional neural networks. *Advances in Neural Information Processing Systems, 25*, 1097–1105.
8. Devlin, J., Chang, M. W., Lee, K., & Toutanova, K. (2019). BERT: Pre-training of deep bidirectional transformers for language understanding. In *Proceedings of the 2019 Conference of the North American Chapter of the Association for Computational Linguistics: Human Language Technologies, Volume 1 (Long and Short Papers)* Association for Computational Linguistics, 4171–4186.
9. Devlin, J., Chang, M. W., Lee, K., & Toutanova, K. (2018). BERT: Pre-training of deep bidirectional transformers for language understanding. arXiv preprint arXiv:1810.04805. https://arxiv.org/abs/1810.04805.
10. Géron, A. (2019). *Hands-On Machine Learning with Scikit-Learn, Keras, and Tensor-Flow: Concepts, Tools, and Techniques to Build Intelligent Systems.* O'Reilly Media.
11. Vaswani, A., Shazeer, N., Parmar, N., Uszkoreit, J., Jones, L., Gomez, A. N., & Polosukhin, I. (2017). Attention is all you need. *Advances in Neural Information Processing Systems*, 30, 5998–6008. https://papers.nips.cc/paper/7181-attention-is-all-you-need.
12. Esteva, A., Kuprel, B., Novoa, R. A., Ko, J., Swetter, S. M., Blau, H. M., & Thrun, S. (2017). Dermatologist-level classification of skin cancer with deep neural networks. *Nature*, 542(7639), 115–118.
13. Gómez-Bombarelli, R., Wei, J. N., Duvenaud, D., Hernández-Lobato, J. M., Sánchez-Lengeling, B., Sheberla, D., & Aspuru-Guzik, A. (2018). Automatic chemical design using a data-driven continuous representation of molecules. *ACS Central Science*, 4(2), 268–276.
14. Bojarski, M., Yeres, P., Choromanska, A., Choromanski, K., Firner, B., Jackel, L. D., & Muller, U. (2016). *End to end learning for self-driving cars.* arXiv preprint arXiv:1604.07316. https://arxiv.org/abs/1604.07316.
15. Brown, N., Balasubramanian, V. N., Caurin, G. A., Huang, E. H., Minhas, A., Mukherjee, A., & West, A. D. (2018). Renewable energy optimization in the age of AI. *IEEE Power and Energy Magazine*, 16(3), 53–62.
16. Chen, L., Mukherjee, A., Zhang, H., & Ji, S. (2018). Efficient classification for additive manufacturing data: A machine learning approach. *Procedia CIRP*, 74, 98–102.
17. Zemel, R., Wu, Y., Swersky, K., Pitassi, T., & Dwork, C. (2013). *Learning fair representations.* In *Proceedings of the 30th International Conference on Machine Learning* (Vol. 28, pp. 325–333). http://proceedings.mlr.press/v28/zemel13.html.

18. Ribeiro, M. T., Singh, S., & Guestrin, C. (2016). *"Why should I trust you?" Explaining the predictions of any classifier.* In *Proceedings of the 22nd ACM SIGKDD International Conference on Knowledge Discovery and Data Mining* (pp. 1135–1144). http://dl.acm.org/doi/10.1145/2939672.2939778.

19. Shvachko, K., Kuang, H., Radia, S., & Chansler, R. (2010). *The Hadoop distributed file system.* In *Proceedings of the 2010 IEEE 26th Symposium on Mass Storage Systems and Technologies (MSST)* (pp. 1–10). http://ieeexplore.ieee.org/document/5496794.

20. Armbrust, M., Fox, A., Griffith, R., Joseph, A. D., Katz, R., Konwinski, A., & Zaharia, M. (2010). A view of cloud computing. *Communications of the ACM*, 53(4), 50–58.

21. Zaharia, M., Chowdhury, M., Das, T., Dave, A., Ma, J., McCauley, M., & Stoica, I. (2010). Resilient distributed datasets: A fault-tolerant abstraction for in-memory cluster computing. In *Proceedings of the 9th USENIX Conference on Networked Systems Design and Implementation* (pp. 2–2).

22. Dilsizian, S. E., & Siegel, E. L. (2014). Artificial intelligence in medicine and cardiac imaging: Harnessing big data and advanced computing to provide personalized medical diagnosis and treatment. *Current Cardiology Reports*, 16(1), 441.

23. Zheng, R., Wang, H., Liu, J., & Li, J. (2017). *Big data-driven supply chain management: A conceptual framework.* In *Proceedings of the 2017 International Conference on Service Systems and Service Management* (pp. 1–6). https://ieeexplore.ieee.org/document/7995746.

24. Kumar, M., Patel, A., Shah, N., & Verma, A. (2018). Fraud detection in financial data using big data analytics. *Procedia Computer Science*, 132, 1576–1585.

25. Calabrese, F., Di Lorenzo, G., Liu, L., & Ratti, C. (2010). Estimating origin-destination flows using mobile phone location data. *IEEE Pervasive Computing*, 9(4), 36–44.

26. Jagadish, H. V., Gehrke, J., Labrinidis, A., Papakonstantinou, Y., Patel, J. M., Ramakrishnan, R., & Shahabi, C. (2014). Big data and its technical challenges. *Communications of the ACM*, 57(7), 86–94.

27. Kitchin, R. (2014). *The Data Revolution: Big Data, Open Data, Data Infrastructures and Their Consequences.* Sage.

28. Dettori, A., & Santana, J. (2019). *Blockchain-based solution for document integrity and digital preservation.* In *2019 IEEE International Conference on Blockchain and Cryptocurrency* (pp. 1–8). Miami, FL, USA. https://ieeexplore.ieee.org/document/8839168.

29. Chen, M. C., & Hsieh, C. L. (2019). A decentralized platform for enhancing trust in legal information. In *2019 IEEE International Congress on Big Data* (pp. 1–8). Milan, Italy. https://ieeexplore.ieee.org/document/8886321.

30. Ramachandran, M., Ranjan, R., & Khan, S. U. (2020). *Smart contracts: Bridging the gap between expectations and reality.* In *2020 IEEE International Conference on Blockchain and Cryptocurrency* (pp. 1–7). Toronto, ON, Canada. https://ieeexplore.ieee.org/document/9157731.

31. Ahram, T., & Al-Turjman, F. (2019). *Blockchain for Intellectual Property Rights: A Survey.* IEEE Access.

32. Lin, Y., Zhang, Y., & Wu, X. (2020). *Blockchain-enabled E-discovery and data privacy.* In *2020 IEEE International Conference on Blockchain and Cryptocurrency* (pp. 1–6). Toronto, ON, Canada. https://ieeexplore.ieee.org/document/9162550.

33. Smith, A., & Johnson, B. (2020). Artificial intelligence in legal document review. *Journal of Legal Technology*. https://www.jltjournal.org/article/view/123456

34. Li, Y., & Chen, W. (2019). Automated legal document analysis using AI techniques. *International Journal of Legal Studies and Research*. https://www.ijlsr.org/article/view/789012

35. Liu, Y., Wang, F., & Wang, X. (2018). *AI-enabled legal research: Harnessing natural language processing. Journal of Legal Information Management.* https://www.jlimjournal.org/article/view/345678

36. Choi, J., Kim, S., & Park, J. (2021). *Predicting case outcomes using machine learning techniques*. *Journal of AI and Law*. https://www.jailjournal.org/article/view/901234
37. Ramirez, C., & Cox, L. (2019). Automating contract analysis with artificial intelligence. *Artificial Intelligence for Legal Professionals*. https://www.aiflprojournal.org/article/view/567890.
38. D'Ignazio, J., & Napoli, P. J. (2019). *Making Data and AI Work in the Public Interest: Lessons from Data for Black Lives*. Data & Society Research Institute.
39. Katz, D. M., Bommarito, M. J., & Blackman, J. (2017). A general approach for predicting the behavior of the Supreme Court of the United States. *PLoS One*, 12(4), e0174698.
40. Iqbal, A., & Lim, Y. F. (2016). *Automated extraction and classification of business intelligence data from contracts*. In *2016 IEEE International Conference on Big Data (Big Data)* (pp. 1–7). Washington, DC, USA. https://ieeexplore.ieee.org/document/7840712.
41. Suleyman, A., et al. (2019). Big data, machine learning, and GDPR: Privacy in the age of AI. *European Data Protection Law Review*, 5(1), 29–41.

2 Introduction to Digital Forensics

*Vaishali C. V., Vennila Thirumalaiswamy, and
Muneeswaran Thillaichidambaram*

2.1 INTRODUCTION

A field of study called "digital forensics" studies how to protect systems and the
data they are connected to. It also covers occupation and evidence analysis to determine what led to the attack in the case. The application of modern technology has
advanced the future in a way that it can be predicted by capturing data accurately.
A safeguarding mechanism in place with digital forensics is capable of identifying
and evaluating evidence going forward. A tool for learning that will improve the
system's security and analytical capacity to track information and computer hazards
is called "digital forensics."[1]

The early years of the 21st century saw the emergence of national policies pertaining to digital forensics. Court presentations of digital forensics that are required for
evidence are possible. Within the field of digital forensics, artificial intelligence (AI)
is one of the newest and most rapidly evolving technologies. It has a big influence on
the instruments and procedures used to investigate, track, and display crime scenes
and to create efficient plans of action for fending off threats and attacks occurring
online. Machine-oriented designed work, which can reduce human error and boost
quality while attaining maximum results with fewer mistakes, has taken the position
of human labor in this technology. In the area of digital forensics, it might affect the
result and examine the evidence more successfully to monitor the results. [2]

2.2 TRADITIONAL DIGITAL INVESTIGATION PROCESS

The processing method of the traditional digital investigative method used in many
countries is described later. The digital devices and detectives interested in the data
stored on these devices are displayed on the left and right side, respectively. A digital
investigator might be involved, depending on the nature of the case. An intrusion
detection system may identify an incursion, a victim may report a crime, law enforcement may be contacted to the scene, or a combination of these may initiate a case.
Digital elements are typically included in traditional crimes like arson, child abuse,
and murder, as well as cybercrimes like phishing, denial-of-service assaults, and
hacking. The majority of detectives have little to no experience with digital gadgets
due to their lack of training, so digital investigators answer a specific question related
to the digital devices (see Figure 2.1).

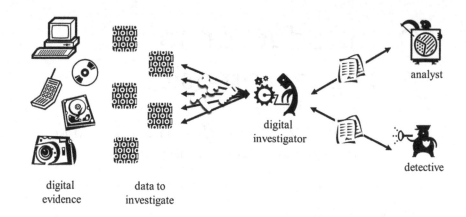

digital
evidence

data to
investigate

digital
investigator

analyst

detective

FIGURE 2.1   The process of conventional digital forensics.

Source: Permission from [3].

Creating forensic copies of the digital devices is usually the initial task (collection and authentication). To recover deleted data, carve unallocated space, extract archives, and other tasks, a variety of tools and scripts can be used. Scripts can be used to connect these routine processes [4]. Digital evidence is made visible during this examination procedure, enabling a detective to review the data. The data may then be indexed, harvested, and organized logically as a potential next step. Since it is not always obvious if a digital device even includes pertinent trace information, the incident might not be identified until the very end of the investigation. In order to ascertain whether an incident, and what kind, took place, the integrated digital forensic process model (IDFPM) depends on the identification phase of the digital inquiry. It is hard to undertake classification, organization, and comparison without identification. In criminal cases, it is more typical for a detective to create inquiries based on case data that isn't digital. Digital investigators provide answers to these queries. Before obtaining pertinent information or posing a follow-up query to clarify findings or narrow down the pool of results, the investigator must sort through these results. Any stages that were skipped before could be completed to recreate the official report if pertinent information is discovered [3].

2.3   DIGITAL FORENSICS PROCESS

Digital forensics is able to gather and examine digital evidence in its original format from a variety of devices, including hardware, disk drives, USBs, gateways, mobile devices, and PCs. Since authority and control in traditional forensics can be governmental, organizational, or individual, they are typically shared on an electronic or standard format [5]. The digital forensics process is shown in Figure 2.2.

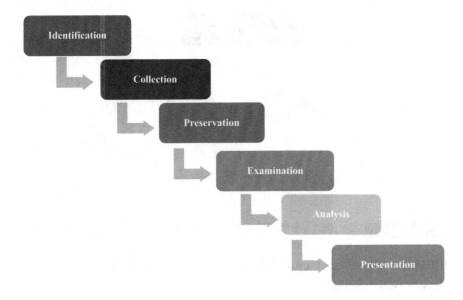

FIGURE 2.2 Digital forensics process.

Source: Adapted from [6].

2.3.1 PROCESS OF DIGITAL FORENSICS

Digital forensics entails the following steps [7]:

- *Identification.* Identification of the primary purpose of an investigation and the resources required to accomplish it.
- *Preservation.* Data isolation, securing, and preservation.
- *Analysis.* Identification of the tools and techniques required to process data and to analyze and interpret the results.
- *Documentation.* Documentation of the crime scene with photograph, sketches, and crime scene mapping.
- *Presentation.* Process of summarization and explanation for conclusion [7].

In "Integrating Forensic Techniques" (SP 800–86), the US National Institute of Standards and Technology suggested a four-phase paradigm for digital forensics in 2006 (see Figure 2.3). The phases of digital forensics investigation are as follows: collection in identifying evidence at the scene, labeling, documenting, and finally, gathering it. *Examination* entails determining which forensic tools and techniques to use to extract pertinent digital evidence while maintaining its integrity, analysis involves evaluating the extracted evidence to ascertain its applicability and usefulness to the case, and reporting includes the actions taken during the process and the presentation of the results [8].

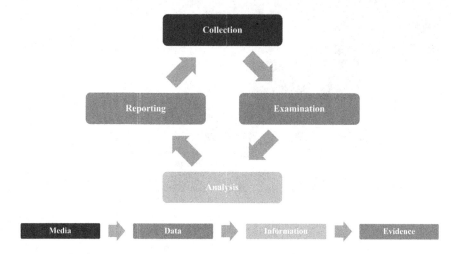

FIGURE 2.3   Four-phase life cycle of digital forensics investigation model.

Source: Adapted from [6].

2.4   ARTIFICIAL INTELLIGENCE

A significant and well-established field in contemporary computer science, artificial intelligence (AI) frequently offers a way to solve computationally difficult or large-scale issues in a reasonable amount of time. The creation of computer programs which can carry out operations that shall be considered as an ordinary call in terms of human intelligence is referred to as *artificial intelligence*. In order to give machines the ability to reason, learn from data, and make decisions, some algorithms and models must be developed. Since it was first developed, artificial intelligence has advanced greatly. At first, artificial intelligence was restricted to rule-based systems that carried out preset commands. On the other hand, AI systems can now learn from enormous volumes of data and gradually improve their performance thanks to developments in machine learning and deep learning [9]. In computing, the field of "digital forensics" is growing in importance and frequently calls for the deft examination of vast quantities of intricate data. AI thus appears to be the perfect solution to address a large number of the issues that digital forensics is now facing [10].

Digital forensics is essential to the investigation of crimes and the administration of justice in the modern digital era. Because technology and the internet are used by everyone, thieves have discovered new ways to use digital systems for their illegal activities. It is the responsibility of digital forensics specialists to locate and examine these digital traces that offenders leave behind in order to compile a case. To extract and analyze digital evidence, a variety of methods and instruments is used in digital forensics. These include data recovery, which aims to retrieve lost or buried data that might contain important information, and forensic imaging, which makes an exact replica of a digital device to maintain its state at the moment of seizure [11].

2.4.1   AI Is Revolutionizing Digital Forensics

Digital forensics has undergone a revolution because of AI tools like machine learning and pattern recognition, which automate the processing of digital evidence. Investigators can uncover suspicious activity or hidden information by using machine learning algorithms that have been trained on massive datasets to discover patterns and abnormalities in digital data. Natural language processing (NLP) methods have been used in data recovery and analysis in addition to machine learning. Using unstructured data, such as emails or text messages, natural language processing (NLP) enables investigators to retrieve pertinent information that can be used to reconstruct conversations or reveal covert communication.

2.4.2   Challenges and Opportunities at the Intersection

Although artificial intelligence (AI) has great potential for digital forensics, there are drawbacks and moral dilemmas. AI models' interpretability is one of the difficulties. The complexity of AI systems makes it harder to comprehend how they make decisions, which raises questions about the accuracy and transparency of their conclusions. The constant change in technology presents another difficulty. Forensic investigators are forced to constantly modify their methodologies in order to stay up with the rapid changes in technology, including new devices, software upgrades, and encryption techniques. Notwithstanding these difficulties, the field of digital forensics and artificial intelligence has a lot of fascinating prospects. Artificial intelligence (AI) can help find pertinent evidence, automate tedious tasks, and lighten the burden on investigators.

2.5   AI TECHNIQUES IN DIGITAL FORENSICS

Digital forensics uses a variety of AI techniques to increase the effectiveness and precision of investigations (see Figure 2.4).

2.5.1   Pattern Recognition and Machine Learning

Digital forensics has used machine learning techniques, like support vector machines and neural networks, to categorize and examine various kinds of digital evidence. These algorithms enable researchers to extract significant insights from complicated datasets by seeing patterns in data and forecasting outcomes based on past performance. Digital forensics has also found use for pattern recognition techniques, like speech and picture recognition. By using these techniques, investigators can more thoroughly understand the events they are looking into by identifying and analyzing visual or aural evidence, such as recorded conversations or surveillance footage.

2.5.2   Natural Language Processing in Data Recovery

Through the use of natural language processing techniques, investigators can glean valuable information from textual material. Investigators can use methods like text mining and sentiment analysis to extract pertinent data from chat logs, social media

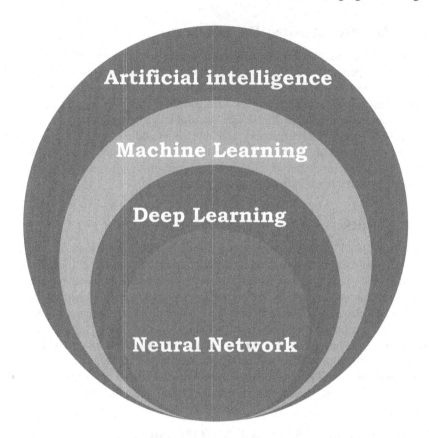

FIGURE 2.4   Different design points of five levels of artificial intelligence.

Source: Adapted from [12].

posts, and emails. Reconstructing discussions, identifying important players, and comprehending the background of the investigation are all made easier by this.

2.6   AI'S ROLE IN DIGITAL FORENSICS IN THE FUTURE

Promising advancements and obstacles that must be overcome are presented by the use of AI in digital forensics.

2.6.1   Forecasting Developments and Trends

AI is predicted to become more important in digital forensics as it develops. It is possible to anticipate changing dangers, assess new technologies, and improve the effectiveness of investigations by utilizing predictive analytics and machine learning models. Artificial intelligence (AI) can also help automate the tiresome and time-consuming processes of gathering and analyzing evidence, freeing up investigators to concentrate on more complex analysis and decision-making.

2.6.2 ETHICS-RELATED CONCERNS AND IMPLICATIONS

There are possible ramifications and ethical questions raised by the growing use of AI in digital forensics. In order to preserve the integrity of investigations and avoid any biases or unjust findings, it is imperative to guarantee transparency and accountability when using AI algorithms. Concerns about privacy also surface when AI is used in digital forensics. Finding the ideal balance between the right to privacy and the necessity of thorough investigations is a difficult issue that calls for careful thought and strong security measures.

2.7 CHALLENGES IN DIGITAL FORENSICS

A variety of difficulties, including ethical, legal, and technical ones, might arise during digital forensics investigations. The volume and complexity of digital data, along with the quick advancement of technology, present technical hurdles. Admissibility of digital evidence in court is one legal difficulty; privacy and personal data protection are two ethical challenges. The rapidly evolving nature of technology is a problem for professionals in the field of digital forensics. Professionals in digital forensics may find it challenging to stay current with new methods and instruments due to the quick speed of technological advancement.

Organizations can, however, overcome this difficulty by funding their digital forensics experts' continual training and development initiatives. These courses may guarantee that professionals have the knowledge needed to properly examine such events and also keep them abreast of the most recent developments and practices in the industry. The admissibility of digital evidence in court is another difficulty for those working in digital forensics. Digital evidence needs to be gathered and stored in a way that guarantees its integrity and authenticity in order for it to be accepted. This can be difficult because digital data is quickly removed or changed. To guarantee that digital evidence is accurately gathered and kept, digital forensics specialists must adhere to established protocols.

Digital forensics also faces ethical challenges related to privacy and personal data protection. The requirement to preserve private information and personal data must be balanced with the necessity for digital forensics specialists to look into such events. Ensuring that investigations are carried out in an ethical and responsible manner, together with putting in place the proper policies and processes for handling personal data, can be part of this [13].

2.8 CONCLUSION

The study of digital forensics is an ongoing field that requires constant modifications to remain effective. Changes should be implemented appropriately and at the appropriate technological peak. The application must be durable, capable of running on multiple platforms, and independent in order to improve its quality. The application of artificial intelligence to digital forensics has the potential to significantly and visibly alter the security landscape. Artificial intelligence will support digital forensics technologies in analyzing evidence and simplify jobs to forensics professional experts in analyzing data and drawing appropriate conclusions to determine the outcome of the crime scene. It might also be beneficial to pre-analyze security threats using historical data and to store threat recordings in order to keep the system updated for later usage [14]. The system can be observed and treated with potential modifications and fixes through the use of forensics tools. Digital forensics has undergone a revolution thanks to the application of artificial intelligence, which makes it possible for investigators to efficiently handle and analyze massive amounts of digital data. Artificial intelligence (AI) methods, such as machine learning and natural language processing, provide strong instruments for finding hidden evidence and spotting trends in large complicated datasets. Nonetheless, issues like interpretability and rapidly advancing technology need to be dealt with, and ethical dilemmas need to be carefully handled. Although artificial intelligence (AI) in digital forensics has enormous promise to improve investigations, its application must be ethical and responsible.

REFERENCES

1. Ali, M. R. (2020, December). Digital forensics and artificial intelligence a study. *International Journal of Innovative Science and Research Technology (IJISRT)*, 5(12), 651–654. www.ijisrt.com; ISSN: 2456-2165.
2. Verma, R., Garg, S., Kumar, K., Gupta, G., Salehi, W., Pareek, P. K., & Kniežova, J. (2023, February). New approach of artificial intelligence in digital forensic investigation: A literature review. In *International Conference on Advances in Communication Technology and Computer Engineering* (pp. 399–409). Cham: Springer Nature Switzerland.
3. van Baar, R. B., van Beek, H. M., & Van Eijk, E. J. (2014). Digital forensics as a service: A game changer. *Digital Investigation*, *11*, S54–S62.
4. Carrier, B., & Spafford, E. H. (2003). Getting physical with the digital investigation process. *International Journal of Digital Evidence*, 2(2), 1–20.
5. Alazab, A., Khraisat, A., & Singh, S. (2023). *A Review on the Internet of Things (IoT) Forensics: Challenges, Techniques, and Evaluation of Digital Forensic Tools*. UK: IntechOpen. DOI 10.5772/intechopen.109840.
6. da Silveira, C. M., de Sousa Jr, T. R., de Oliveira Albuquerque, R., Amvame Nze, G. D., de Oliveira Júnior, G. A., Sandoval Orozco, A. L., & García Villalba, L. J. (2020). Methodology for forensics data reconstruction on mobile devices with Android operating system applying in-system programming and combination firmware. *Applied Sciences*, *10*(12), 4231.
7. Rafique, M., & Khan, M. N. A. (2013). Exploring static and live digital forensics: Methods, practices and tools. *International Journal of Scientific & Engineering Research*, 4(10), 1048–1056.
8. Kent, K., Chevalier, S., & Grance, T. (2006). Guide to integrating forensic techniques into incident. *Technical Reports*, 800–886.
9. Patel, R. B. (2023, June 12). *The Use of Artificial Intelligence in Digital Forensics*. Authorea Preprints. *TechRxiv*. DOI: 10.36227/techrxiv.23461142.v1.
10. Mitchell, F. (2010). The use of Artificial Intelligence in digital forensics: An introduction. *Digital Evidence and Electronic Signature Law Review*, 7, 35.
11. Dunsin, D., Ghanem, M. C., & Quazzane, K. (2022). The use of artificial intelligence in digital forensics and incident response in a constrained environment. *International Journal of Information and Communication Engineering*, *16*(8), 280–285.

12. www.constellationr.com/blog-news/mondays-musings-designing-five-pillars-level-1-artificial-intelligence-ethics (last accessed on 20-Apr-24).
13. Duce, D., Mitchell, F., Turner, P., Haggerty, J., & Merabti, M. (2007, July). Digital forensics: Challenges and opportunities. In *2nd Conference on Advances in Computer Security and Forensics (ACSF)'*, LJMU, Liverpool, UK.
14. Kohn, M. D., Eloff, M. M., & Eloff, J. H. (2013). Integrated digital forensic process model. *Computers & Security*, *38*, 103–115.

3 AI in Digital Forensics

Lakshmi K. S.

3.1 INTRODUCTION TO DIGITAL FORENSICS

Digital forensics plays a pivotal role in contemporary investigations, serving as a specialized discipline within forensic science. Its core focus encompasses the identification, acquisition, processing, analysis, and documentation of electronically stored data. Given the pervasive presence of electronic evidence, the expertise of digital forensics professionals is indispensable in law enforcement investigations. From computers and smartphones to remote storage, unmanned aerial systems, and maritime equipment, a diverse array of devices may contain critical electronic evidence. The principal aim of digital forensics is to extract pertinent information from such evidence, thereby converting it into actionable intelligence that can be utilized for prosecution purposes. Each phase of the process adheres to stringent forensic methodologies, ensuring the admissibility of findings in judicial proceedings.

3.2 THE PHASES OF A DIGITAL FORENSICS INVESTIGATION

Stage 1: Identification

The initial phase of a digital forensics investigation involves identifying relevant digital devices and data sources. This encompasses both organizational assets, like computers, and user-owned devices, such as smartphones and tablets. Seizing and isolating these devices is crucial to prevent any potential tampering. In cases involving servers, networks, or cloud storage, the investigator or organization must secure access and restrict unauthorized modification.

Stage 2: Extraction and Preservation

Following the secure acquisition of relevant devices, forensic specialists utilize specialized techniques to extract potentially relevant data. This process often involves creating a forensic image, a meticulous digital copy of the original data. This copy serves as the primary source for analysis, while the original data and devices remain securely stored, ensuring their integrity even if the investigation encounters unforeseen challenges.

Stage 3: Analysis

After securing and isolating targeted devices and creating a secure copy of the data, digital forensic examiners employ various techniques to extract and analyze relevant information. This stage primarily focuses on identifying potential evidence of wrongdoing, which may involve:

- *Reverse steganography.* Uncovering hidden data embedded within seemingly innocuous files like images, leveraging specialized techniques to uncover hidden patterns or character sequences.

DOI: 10.1201/9781003501152-3

- *File carving.* Recovering deleted files by searching for fragmented remnants left behind after deletion attempts.
- *Keyword searching.* Employing specific keywords to locate and analyze information pertinent to the investigation, even within deleted data.

These represent a subset of the various techniques utilized by digital forensics professionals during their investigative efforts.

Stage 4: Documentation

Following the analysis phase, the investigator meticulously documents the entire investigative process and its findings. This comprehensive documentation serves as a visual representation of the investigation's timeline, including the chronology of events related to potential wrongdoing, such as data breaches, financial misconduct, or network intrusions.

Stage 5: Presentation

Upon completion of the investigation, the findings are presented to relevant authorities, such as courts or internal investigation committees. Digital forensics investigators may also serve as expert witnesses, providing summaries of the discovered evidence and confidently presenting their findings clearly and concisely.

3.3   AI IN DIGITAL FORENSICS: A POTENTIAL GAME-CHANGER

As the digital environment undergoes perpetual evolution, criminal activities within it also adapt and evolve. The realm of cybercrime is undergoing substantial growth, necessitating more resilient and effective investigative approaches. In this context, artificial intelligence (AI) emerges as a promising asset, providing significant advancements in the domain of digital forensics.

Digital forensics encompasses the investigation of digital devices and cyber networks to unearth and interpret electronic evidence. The primary goal lies in preserving evidence in its original state while conducting a structured investigative process. Integrating AI into this field equips specialists with a powerful instrument to navigate increasingly intricate and complex cybercrime scenarios.

The ever-expanding digital landscape has facilitated the rise of sophisticated cybercrime techniques, posing significant challenges to traditional detection and prevention efforts. This is where AI offers a critical solution. Through machine learning, a subfield of AI, systems gain the ability to learn and improve based on experience. These machine learning algorithms are adept at identifying patterns within vast datasets—patterns that might escape human detection or require extensive time to discover.

3.4   REAL-WORLD EXAMPLES: AI STEPPING UP IN DIGITAL FORENSICS

Exploring concrete examples across different subfields is valuable for gaining a deeper understanding of how AI contributes to the digital forensics landscape.

1. Network Forensics

   Case Study: In 2019, Awake Security, a software company, showcased the potential of AI in network forensics through its AI-powered security platform. This platform successfully identified and thwarted a potential data breach attempt by "Fxmsp," a known cybercriminal targeting major antivirus companies. The platform's effectiveness stemmed from its ability to analyze network traffic and leverage AI to detect anomalous behavior, ultimately foiling the cyberattack.

2. Image and Video Forensics

   Application: AI is making significant strides in the field of image and video forensics. For example, Truepic, a technology company striving to authenticate digital photos and videos, utilizes AI to detect "deepfakes"—manipulated media created using AI to depict fabricated scenarios. Truepic's platform analyzes various parameters, including lighting and potential 3D face mask overlays, serving as a powerful tool for detecting image and video manipulation.

3. Cryptocurrency Forensics

   Examples: CipherTrace and Elliptic are leading companies leveraging AI in cryptocurrency forensics to combat illicit activities within the digital currency ecosystem. These platforms employ machine learning technologies to identify suspicious patterns within cryptocurrency transaction chains, supporting investigations and mitigating risks associated with cryptocurrency investments.

3.5   DIVING INTO THE DETAILS: USING AI IN DIGITAL FORENSICS

AI's inherent abilities contribute significantly to its success in digital forensics, specifically through:

1. Pattern Recognition

   AI excels at learning and identifying patterns within vast and intricate datasets. For instance, an AI system can be trained to recognize indicators of potential phishing attempts in emails, such as specific formats, structures, or commonly used phrases by cybercriminals.

2. Anomaly Detection

   This refers to the identification of outliers or unusual data points in datasets. In cybersecurity, this translates to pinpointing any atypical behavior deviating from the norm that could potentially represent a security threat.

3. Predictive Analysis

   Employing statistical algorithms and machine learning techniques, this approach aims to assess the probability of forthcoming events. In the realm of digital forensics, predictive models can scrutinize past attack patterns and predict potential future assaults, empowering the implementation of proactive security measures.

4. Automated Log Analysis
   Security teams often face a significant volume of log files originating from various systems, apps, and network devices. Manual examination of these logs can prove laborious and prone to errors. AI-driven log analysis presents a remedy.

   AI algorithms demonstrate proficiency in handling extensive volumes of log files and systematically reviewing them to discern patterns and anomalies. This advancement augments the efficiency and precision of log analysis, enabling investigators to direct their attention toward pertinent areas while circumventing the need for manual scrutiny.

5. Malware Detection
   Malware is evolving so quickly that sophisticated detection techniques are required. AI-driven platforms make use of machine learning to:
   • Analyze and scan code
   • Learn user behavior patterns
   • Detect malicious software more effectively
   • Assist in removing malware from compromised systems
   Security firms utilize AI algorithms to continuously learn from known malware samples and their attributes. By training these algorithms on comprehensive datasets, they can identify and categorize new and previously unidentified malware variants, preemptively identifying potential attacks.

6. Image and Video Analysis
   The examination of digital images and videos holds significant importance in the realm of digital forensics. AI algorithms are capable of:
   • Sorting through extensive quantities of multimedia content
   • Rapidly detecting faces, objects, or text present within images and videos
   • Substantially expediting the process of locating and extracting vital evidence
   For example, facial recognition technology powered by AI can efficiently analyze large volumes of video footage, accurately identifying individuals of interest within crowded environments and streamlining the identification process.

7. Natural Language Processing (NLP)
   AI technologies such as natural language processing (NLP) facilitate the extraction of relevant insights from extensive text datasets. Textual data, such as emails, chat transcripts, and documents, frequently harbor significant evidence in digital inquiries. AI offers enhanced efficiency and accuracy in:
   • Discerning relationships
   • Detecting patterns
   • Identifying crucial individuals in text-centric investigations
   By leveraging AI-powered NLP algorithms, text data can be comprehensively processed and analyzed, enabling the identification of recurring phrases, suspicious trends, and correlations among individuals. This accelerates the investigative process and facilitates timely interventions.

8. Analyzing Network Traffic

The identification and mitigation of cyberattacks depend heavily on the monitoring and analysis of network traffic patterns. AI capabilities encompass:

- Automatically training algorithms to analyze network packets
- Discerning deviations from typical traffic patterns
- Prompting alerts when anomalies warrant additional scrutiny
- Correlating network events with established attack patterns to furnish incident response teams with invaluable insights

9. Forensic Triage

Large amounts of data are involved in digital investigations, so investigators must act quickly to rank the most important evidence. Large volumes of digital material are categorized and classified by machine learning algorithms according to how relevant they are to a particular inquiry. In order to prioritize files, these technologies examine file information, content, and other characteristics. They also continuously learn to identify significant material more accurately. Artificial intelligence (AI) enables forensics teams to rapidly locate and concentrate on the most crucial evidence, resulting in quicker and more efficient investigations while optimizing the use of available resources.

3.6 CHALLENGES DIGITAL FORENSIC SPECIALISTS FACE

Digital forensic analysts face a multifaceted and ever-evolving landscape, with three key challenges demanding solutions:

1. Data Diversity

Investigations often involve evidence from diverse sources, including mobile phones, computers, laptops, tablets, cameras, and storage devices. Each device can have unique operating systems that conceal digital evidence in different ways.

Challenge: Collecting and analyzing this multifaceted data can be time-consuming and labor-intensive, potentially hindering the investigation's progress.

2. Scalability

The current capabilities of digital forensics technology often struggle with the scalability required to efficiently manage the vast amount of data generated by multiple ongoing investigations.

Challenge: This lack of scalability can lead to a backlog of unanalyzed cases, potentially jeopardizing the integrity of crucial evidence due to delays in examination.

3. Limited Staffing

The field of digital forensics frequently faces staffing shortages, further complicating the process of collecting and analyzing evidence effectively.

Challenge: When investigators are overburdened, crucial evidence might be overlooked or lost, potentially compromising the integrity of the investigation.

3.6.1 AI AS A POTENTIAL SOLUTION

The aforementioned challenges highlight a significant opportunity for AI integration within the field of digital forensics. AI presents the potential to:

Automate data collection and analysis. Streamlining the processing of diverse data types from various sources, thereby expediting investigations.

Enhance scalability. Providing the capability to efficiently manage and analyze large volumes of data, potentially eliminating backlogs and ensuring timely evidence examination.

Augment limited staffing. By automating repetitive tasks and providing intelligent data analysis, AI can support and empower existing forensic teams, enabling them to maximize their efficiency and effectiveness despite staffing limitations.

3.7 AI-DRIVEN ADVANCEMENTS IN DIGITAL FORENSICS RESEARCH

The realm of digital forensics has undergone rapid expansion, propelled by the expanding role of technology in criminal endeavors and the consequential elevation of digital evidence in investigative processes. Nonetheless, obstacles endure, including the heterogeneous nature of digital data, the scalability requirements for processing vast datasets, and the constraints faced by inadequately staffed teams. Artificial intelligence (AI) and machine learning (ML) emerge as auspicious remedies to these challenges, providing automated data management capabilities, enhanced scalability for handling extensive data volumes, and bolstered support for teams with limited personnel resources.

Walker et al. [1] report that the field of digital forensics has grown rapidly, employing technology to gather and analyze digital evidence for use in criminal investigations. The increasing importance of digital evidence means that effective methods for examining criminal activity are imperative. Garfinkel et al. [2] assert that AI and ML are crucial instruments in the field of digital forensics, which analyzes vast amounts of digital data in order to look into criminal cases. They assist with challenging data processing tasks that are beyond the capabilities of humans. There are still problems, like the application incompatibility that Cabitza et al. [3] mentioned, despite the tremendous advancements.

Furthermore, privacy restrictions create ethical and legal challenges when collecting and reconstructing data with the intention of identifying criminals. In the field of digital forensics, more versatile and effective tools must be developed to meet these challenges and remain competitive with offenders.

Dunsin et al. [4] have explored various applications of AI and ML in digital forensics, encompassing a wide range of areas, as shown in Figure 3.1. These include automated investigation models [5], using data mining techniques for analyzing digital evidence [6], methods for automated malware detection using memory image analysis [7], and the use of neural networks to reconstruct events and retrieve crucial information from digital forensics data [8]. Additionally, research has addressed challenges,

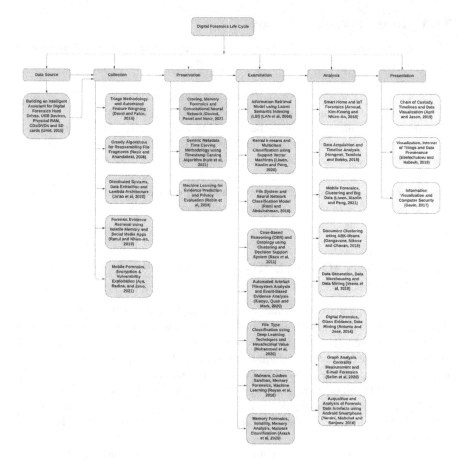

FIGURE 3.1 Systematic literature survey conducted by Dunsin et al.

Source: Adapted from [4].

such as analyzing alterations in storage devices for forensic purposes [9], developing tools for visualizing the chain of custody process in criminal investigations [10], and extracting memory dumps from various platforms to acquire forensic evidence, particularly from social media and instant messaging applications [11]. Furthermore, studies have explored the use of AI for case prioritization [12, 13], large-scale email analysis [14], and big data–driven investigations [15]. Advancements have also been made in data carving and memory forensics with the development of tools like FiFTy [16] and methods for detecting audio forgeries [17].

As the field of digital forensics evolves, continued research and development in AI and ML integration are crucial to maintaining pace with the changing landscape. Addressing concerns regarding data validity, developing AI systems tailored to specific regions, and defining clear objectives for AI and ML implementation will be essential for ensuring successful adoption in digital forensics practices.

3.8 AI CHALLENGES IN DIGITAL FORENSICS

A recent analysis by Johannes et al. [18] explored the potential of strong AI in digital forensics and identified several key challenges associated with its implementation. These challenges highlight the need for careful consideration before widespread adoption.

Explainability [19]. In forensic investigations, evidence needs to be presented in court and understood by non-experts. When using AI models, ensuring explainability is crucial. This involves transparency in how algorithms and models arrive at their conclusions, allowing users to understand the reasoning behind the results.

Human oversight [20]. Current AI models may not consistently achieve the accuracy required for forensic investigations. To address this, one can implement a "human-in-the-loop" approach as a mitigation strategy.

Here, humans supervise the AI's output and intervene when necessary. Confidence estimates provided by some models can help identify situations requiring human intervention, reducing unnecessary reviews.

Black box models and confidence intervals [21]. Some AI models are opaque, making it difficult to understand their inner workings. This opacity can be addressed by incorporating human oversight and through the development of methods to generate confidence intervals for the model's output. Additionally, human involvement allows for:

- *Error correction.* Humans can identify and correct errors before relying on the model's output for decision-making.
- *Improved training data.* By identifying errors, humans can annotate new training data, allowing the model to learn and improve over time.
- *Adaptability.* Humans can adapt model parameters to specific situations, enabling the use of different models for different tasks, such as language selection for automated translation.

Adversarial models [22]. AI models trained on specific data can be vulnerable to adversarial attacks. These attacks involve crafting inputs designed to manipulate the model's output, potentially leading to erroneous results. The existence of such vulnerabilities necessitates careful consideration of the potential for misuse and the development of robust defense mechanisms.

Bias in data and models [23, 24]. Bias in training data can lead to biased outputs from AI models. Ethical considerations require acknowledging and mitigating such biases. Additionally, any bias present in the model's results should be documented and transparently communicated when presenting forensic evidence.

Model sharing [25]. Sharing pretrained AI models can be beneficial as it reduces training effort for new users. However, this practice can pose security risks, such as the potential for extracting private information from the model itself or the creation of white box adversarial attacks.

Autonomy [26]. As AI systems become increasingly autonomous, concerns arise regarding their ability to make independent decisions. This autonomy introduces the risk of suboptimal choices and raises additional challenges in explaining the reasoning behind such decisions, especially if they are not solely data-driven but also influenced by contextual factors.

Consciousness and conscience [27]. If AI systems were to evolve to a level of consciousness and conscience, ethical considerations regarding their decision-making processes would become paramount. Meissner proposes a hierarchy of such properties, highlighting the increasing complexity and potential challenges associated with higher levels of intelligence.

Trust [28, 29]. As AI capabilities advance, it may become increasingly difficult for humans to fully understand the reasoning behind complex AI systems. This raises concerns about trust in the technology and its outputs. Mitigating these concerns will require ongoing efforts to ensure transparency, explainability, and responsible development practices.

While AI holds significant potential for enhancing digital forensics capabilities, the challenges outlined earlier necessitate careful consideration and ongoing research efforts. Addressing these challenges will be crucial to ensure the responsible and effective integration of AI into the field of digital forensics.

3.9   CHALLENGES AND FUTURE DIRECTIONS IN AI-INTEGRATED DIGITAL FORENSICS

The integration of artificial intelligence (AI) into digital forensics, like any technological advancement, presents certain challenges. One such challenge lies in the training of AI models, which often require vast amounts of data. This data may not always be readily available due to privacy concerns. Additionally, potential biases within the training data can manifest as biases in the AI's decision-making, potentially leading to misleading or inaccurate results if left unchecked.

Despite these challenges, the future of AI in digital forensics remains promising. Ongoing research and development efforts aim to mitigate these limitations and enhance the efficiency and effectiveness of AI-powered forensics solutions. As the digital landscape continues to expand and evolve, AI is likely to solidify its position as a crucial tool in the fight against cybercrime, offering valuable assistance to digital forensic investigators.

3.10   MAKING AI WORK FOR DIGITAL FORENSICS: CONCLUSION

Artificial intelligence (AI) presents a significant opportunity to enhance capabilities within the field of digital forensics. Its ability to learn and adapt makes it a valuable ally in the ongoing fight against cybercrime. By analyzing vast quantities of data, identifying anomalies, and continuously learning from these patterns, AI can offer crucial support to digital forensic investigators.

However, it is essential to emphasize that AI tools are not meant to replace human expertise. They function as powerful supplements, augmenting human capabilities during investigations. While AI excels at data processing and pattern recognition, the critical thinking, judgment, and analytical skills of human investigators remain irreplaceable.

In the face of a constantly evolving digital landscape marked by ever-advancing cyber threats, the integration of AI equips forensic investigators with advanced tools to stay ahead. This integration not only provides enhanced capabilities but also presents a new way of approaching problem-solving within the realm of digital forensics. The potential of AI to complement human expertise and contribute to new investigative strategies holds immense promise for the future of digital forensics.

3.11 THE FUTURE OF AI IN DIGITAL FORENSICS

The integration of artificial intelligence (AI) into digital forensics has emerged as a critical area of discussion among experts in the field. This integration holds significant promise for enhanced capabilities, particularly in areas like data analytics, pattern recognition, and anomaly detection. As AI technologies are further developed and deployed, fostering their responsible and ethical use will be paramount.

To achieve this, addressing key challenges is essential. Ensuring the accuracy of AI-driven forensics tools will be crucial, potentially requiring the implementation of multi-factor authentication methodologies, such as fingerprint or iris recognition. Additionally, concerns surrounding data privacy and potential biases within AI algorithms must be carefully considered and mitigated. By proactively addressing these challenges, we can ensure that AI integration not only empowers digital forensics but also upholds ethical and responsible practices.

REFERENCES

1. Walker, C., 2001. Digital evidence and computer crime: Forensic science, computers and the Internet. *Crime Prevention and Community Safety International Journal*, 3, 87–88.
2. Garfinkel, S.L., 2010. Digital forensics research: The next 10 years. *Digital Investigation*, 7, S64–S73. https://doi.org/10.1016/j.diin.2010.05.009.
3. Cabitza, F., Campagner, A., Basile, V., 2023. Toward a perspectivist turn in ground truthing for predictive computing. In: *Proceedings of the AAAI Conference on Artificial Intelligence* (Vol. 37, pp. 6860–6868). https://doi.org/10.1609/aaai.v37i6.25840.
4. Dunsin, D., Ghanem, M., Quazzane, K., 2022. The use of artificial intelligence in digital forensics and incident response in a constrained environment. *International Journal of Information and Communication Engineering*, 16 (8), 280–285.
5. Hasan, R., Raghav, A., Mahmood, S., Hasan, M.A., 2011a. Artificial intelligence-based model for incident response. In: *2011 International Conference on Information Management, Innovation Management, and Industrial Engineering*. https://doi.org/10.1109/iciii.2011.307.
6. Tallon-Ballesteros, A.J., Riquelme, J.C., 2014. Data mining methods applied to a digital forensics task for supervised machine learning. *Computational Intelligence in Digital Forensics: Forensic Investigation and Applications*, 413–428. https://doi.org/10.1007/978-3-319-05885-6_17.
7. Mosli, R., Li, R., Yuan, B., Pan, Y., 2016. Automated malware detection using artifacts in forensic memory images. In: *2016 IEEE Symposium on Technologies for Homeland Security (HST)* (pp. 1–6). IEEE. https://doi.org/10.1109/ths.2016.7568881.
8. Mohammad, R.M., 2018. A neural network based digital forensics classification. In: *2018 IEEE/ACS 15th International Conference on Computer Systems and Applications (AICCSA)*. https://doi.org/10.1109/aiccsa.2018.8612868.

9. Toraskar, T., Bhangale, U., Patil, S., More, N., 2019. Efficient computer forensic analysis using machine learning approaches. In: *2019 IEEE Bombay Section Signature Conference (IBSSC)*. https://doi.org/10.1109/ibssc47189.2019.8973099.

10. Tanner, A., Bruno, J., 2019. Timely: A chain of custody data visualizer. In: *2019 SoutheastCon* (pp. 1–5). IEEE. https://doi.org/10.1109/southeastcon42311.2019.9020497.

11. Thantilage, R.D., Le Khac, N.A., 2019. Framework for the retrieval of social media and instant messaging evidence from volatile memory. In: *2019 18th IEEE International Conference on Trust, Security and Privacy in Computing and Communications/13th IEEE International Conference on Big Data Science and Engineering (TrustCom/BigDataSE)*. https://doi.org/10.1109/trustcom/bigdatase.2019.00070.

12. Du, X., Hargreaves, C., Sheppard, J., Anda, F., Sayakkara, A., Le-Khac, N.-A., Scanlon, M., 2020. SoK: exploring the state of the art and the future potential of artificial intelligence in digital forensic investigation. In: *Proceedings of the 15th International Conference on Availability, Reliability and Security*. https://doi.org/10.1145/3407023.3407068.

13. Du, X., Le, Q., Scanlon, M., 2020. Automated artefact relevancy determination from artefact metadata and associated timeline events. In: *2020 International Conference on Cyber Security and Protection of Digital Services (Cyber Security)* (pp. 1–8). IEEE. https://doi.org/10.1109/cybersecurity49315.2020.9138874.

14. Ozcan, S., Astekin, M., Shashidhar, N.K., Zhou, B., 2020. Centrality and scalability analysis on distributed graph of large-scale E-mail dataset for digital forensics. In: *2020 IEEE International Conference on Big Data (Big Data)* (pp. 2318–2327). IEEE. https://doi.org/10.1109/bigdata50022.2020.9378152.

15. Song, J., Li, J., 2020. A framework for digital forensic investigation of big data. In: *3rd International Conference on Artificial Intelligence and Big Data* (pp. 96–100). Chengdu, China: ICAIBD. https://doi.org/10.1109/ICAIBD49809.2020.9137498.

16. Mittal, G., Korus, P., Memon, N., 2021. FiFTy: Large-scale file fragment type identification using convolutional neural networks. *IEEE Transactions on Information Forensics and Security*, 16, 28–41. https://doi.org/10.1109/tifs.2020.3004266.

17. Su, Z., Li, Mengke, Zhang, G., Wu, Q., Li, Miqing, Zhang, W., Yao, X., 2023. Robust audio copy-move forgery detection using constant Q spectral Sketches and GA-SVM. *IEEE Transactions on Dependable and Secure Computing*, 20, 4016–4031. https://doi.org/10.1109/tdsc.2022.3215280.

18. Faehndrich, J., Honekamp, W., Povalej, R., Rittelmeier, H., Berner, S., Labudde, D., 2023. Digital forensics and strong AI: A structured literature review. *Forensic Science International: Digital Investigation*, 46 (2023), 301617.

19. Sanchez, L., Grajeda, C., Baggili, I., Hall, C., 2019. A practitioner survey exploring the value of forensic tools, ai, filtering, & safer presentation for investigating child sexual abuse material (CSAM). *Digital Investigation*, 29, S124–S142.

20. Nguyen, T.N., Choo, R., 2021. Human-in-the-loop XAI-enabled vulnerability detection, investigation, and mitigation. In: *2021 36th IEEE/ACM International Conference on Automated Software Engineering (ASE)* (pp. 1210–1212). Melbourne, Australia. https://doi.org/10.1109/ASE51524.2021.9678840.

21. Guo, C., Pleiss, G., Sun, Y., Weinberger, K.Q., 2017. On calibration of modern neural networks. In: *Proceedings of the 34th International Conference on Machine Learning. Proceedings of Machine Learning Research* (Vol. 70, pp. 1321–1330). https://proceedings.mlr.press/v70/guo17a.html.

22. Nowroozi, E., Dehghantanha, A., Parizi, R.M., Choo, K.-K.R., 2021. A survey of machine learning techniques in adversarial image forensics. *Computers & Security*, 100, 102092.

23. Meissner, C.A., Kassin, S.M., 2002. "He's guilty!": Investigator bias in judgments of truth and deception. *Law and Human Behavior*, 26 (5), 469–480.

24. Raji, I.D., Smart, A., White, R.N., Mitchell, M., Gebru, T., Hutchinson, B., Smith-Loud, J., Theron, D., Barnes, P., 2020. Closing the AI accountability gap: Defining an end-to-end framework for internal algorithmic auditing. In: *FAT* '20: Proceedings of the 2020 Conference on Fairness, Accountability, and Transparency* (pp. 33–44). https://doi.org/10.1145/3351095.3372873.
25. Veale, M., Binns, R., Edwards, L., 2018. Algorithms that remember: Model inversion attacks and data protection law. *Philosophical Transactions of the Royal Society A: Mathematical, Physical and Engineering Sciences*, 376 (2133), 20180083.
26. Totschnig, W., 2020. Fully autonomous AI. *Science and Engineering Ethics*, 26 (5), 2473–2485.
27. Meissner, G., 2020. Artificial intelligence: Consciousness and conscience. *AI & Society*, 35 (1), 225–235.
28. Marcus, G., Davis, E., 2019. *Rebooting AI: Building Artificial Intelligence We Can Trust.* Vintage.
29. Siau, K., Wang, W., 2018. Building trust in artificial intelligence, machine learning, and robotics. *Cutter Business Technology Journal*, 31 (2), 47–53.

4 Forensic Intelligence
Bridging Science and Technology in the Digital Era

Mahyuddin K. M. Nasution

4.1 INTRODUCTION

The development of science and technology, although currently not in full scientific methodology, supports each other to meet the challenges of change in order to fulfill human life's needs regarding achieving prosperity. [1] One of these sciences is known as digital forensics—also known as computer forensics—as a branch of forensic science whose scientific discussions include the recovery, investigation, examination, and analysis of material found in all digital devices. [2, 3] It is about science which aims to save and clarify data that has and has not been stored in computers or flows in computer networks or digital devices and filters data to increase trust before carrying out the process of turning it into information, and then knowledge, to support decision-making. [4] Digital forensics has grown as a trendy research focus as the digital world is bombarded with data of uncertain quality and also requires the implementation of technology to improve this scientific field. One technology that influences and increasingly determines the daily lives of individuals and society is artificial intelligence (AI). [5] A technology that dynamically continues to grow to study and accumulate natural phenomena and patterns of human behavior in order to achieve harmony in life. [6] For this reason, this chapter intends to reveal all the roles of AI in the world of forensics and its potential in digital forensics. However, considering the study of both AI and forensics in general, it is too broad to summarize the role of AI in the world of forensics by emphasizing it in appropriate keywords.

4.2 A BACKGROUND

Forensics and *artificial intelligence* (AI) are scientific names related to science and technology, respectively. [7, 8] As a science, forensics, which appeared in a reputable database in 1828, [9] is *a science that aims to recognize the effects of touching nature, organisms, or any event on an object or entity.* [10, 11] Touch is also known as treatment. Every treatment according to its level will change the target object so that sometimes it can no longer be recognized properly by the naked (ordinary) eye and therefore requires such knowledge (which is called forensic) to reveal the existence of the object's identity in what has changed. [12] In order to reveal what the changing object actually is, at least methods, equipment, or auxiliary equipment

DOI: 10.1201/9781003501152-4

in the form of technology are required. Then, as a technology, AI, which appeared in reputable database records in 1960, [13] is a scientific way to automatically recognize certain objects through patterns or arrangements of characteristics that have been manually assessed by humans [14]. However, AI follows human attitudes where learning is the subject of study in this chapter.

4.2.1 TOWARD DIGITAL FORENSICS AS SCIENCE

Humans, in navigating their lives, face challenge after challenge. One of these challenges is the importance of tracing evidence of an event that causes disruption to the achievement of human welfare, [15] and humans name their interests in their own language according to the nation's thinking power, [16] but different words to name the same interests may refer to one word, which linguistically comes from the Latin word *forēnsis*, an adjective referring to "3rd declension," meaning "forum, gathering place." [17] Initially, the confirmation of the use of the term *forensic* referred to legal evidence containing arguments from the person accused or the accuser who explained the problem before a forum. Decisions are taken based on the best arguments presented in the forum. [18] So what is currently in effect in various courts or dialogues is the forensic implementation of an event or presentation, where the presentation contains truth values that present scientific support, then forensics, as a science, systematically obtains the knowledge that underlies a decision. [19] So forensics is a science that uses scientific methods and processes to solve crimes.

Forensics as a science experienced development from the medical world, called entomology, [20] which accompanied legal views or was based on regulations related to handling crime. [21] In the past, in order for criminals not to escape punishment, criminal investigations and trials relied on forced confessions and witness statements; it was difficult to obtain evidence of crimes, but from human experience, there are ways to solve this in terms of increasing the quality of decision-making. [22] Various events gave rise to notes, and from these notes emerged the concept of tracing a crime, which became the source of forensics, including carrying out autopsies. Autopsy is a way to protect evidence during the examination process. [23] So forensic development, in this case, is by forensically also recording the notes, where by studying the notes and accumulating them as something standard, [24] it gives rise to theories or postulates that support each other and are integrated into a science called forensics, which is supported by a philosophy of science following general principles of life or common sense and a methodology for implying and implementing them in upholding justice in law and human rights. [25, 26]

History has recorded that certain regions, several countries, and ethnic groups have practiced activities with forensic nuances regarding various events they experienced. Good or bad, events create records in individuals and communities and then become a culture that is passed down from generation to generation, and some of them are recorded in classic books. [27, 28] In general, the information on the note states that forensic activities are carried out by interacting between forensic objects and natural objects or organisms (biological materials). [29] Indirectly, the interaction must be validated physically by revealing geometric measurements or physical properties. [30] For example, strangulation causes neck cartilage fractures.

[31] Another possibility is that the interaction is proven chemically by revealing that there are different chemical traces in the people involved, [32] or specifically that the interaction has psychological effects on a person: [33] dry saliva due to pressure from guilt, [34] for example. In this way, forensic developments are also guided by different sciences, starting from mathematics, natural sciences, especially physics and chemistry, including psychology. But in essence, all fields require forensics, for example, in the world of economics and management, auditing also aims to carry out forensics on records, or traces of fraud in the world of business and commerce. [35] Therefore, when digital systems emerge, inevitably, both theories and methods of forensics adapt to digital objects or treatments, [36] which indirectly gives rise to what is called digital forensics. [37]

Digital involves electronic computer equipment. [38] Digital implementation leads to two different things, firstly, the use of digital to help manual activities, especially those related to computing and known as digitalization, and, secondly, related to digital content, ranging from data and information to text, images, hyperlinks (URL links), documents, web, audio, and video. [39] Directly, the implication is that digital forensics cannot be separated from the influence of computers and their development. [40] Therefore, digital forensics, also known as computer forensics, [41] is a branch of forensics that includes the recovery, investigation, examination, and analysis of material found in computer devices or other digital machines, such as cell phones, or cellular phones. However, with the increasing popularity of computers as a tool to improve human performance with the existence of networks and the internet, [42] inevitably, not only human work but also the ways of crime have changed. Crime uses computers (digital) not only as tools and targets but also as the source of the crime. Therefore, *digital forensics is a science that is developed based on the principle of zeros and ones to be applied in engineering the actual digital evidence for legal problems that involve technology, system, or not.* [43]

4.2.2 AI AS TECHNOLOGY

Computers as executors of the computing process are faster than humans and can calculate the results to be able to give a learning process, even though the learning is actually a device derived from the capabilities of the human brain. [44] Culturally, this idea has been established by the ancient mythical automaton named Talos, [45] but also indirectly by the revealed scriptures, [46] where the term *intelligence* includes "logic," "understanding," "awareness," "learning," "emotion," "reasoning," "planning," "creativity," "procedures," and "solution." All come from the neuron system. [47] Meanwhile, the term *artificial* refers to a natural copy of something that has been made.

The idea of AI is implicit in various programming software, which implements algorithms into computer programs, where the program becomes an illustration of the capabilities of the human brain, which originally simply made calculations carried out by humans faster. [48] In this way, AI was initially stated as knowledge related to the use of machines to act like the human brain. Likewise, the official use of the term *AI* began in the 1950s at the initiative of John McCarthy, [49] who supported the program's ability to write scripts and the computer's ability to play chess. [50] In this way, scientists declare AI as a part of computer science that makes computer machines able to perform tasks like and as humans do. [51] However, AI is increasingly experiencing acceleration in its development as a technology based on the establishment of formal sciences, such as mathematics and statistics, which are spreading to computer science. [52] Moreover, with this experience, humans have found various ways to adopt natural behavior into an intelligent system called AI, which has presented many algorithms for recognizing life patterns, such as genetic algorithms, fish algorithms, bee algorithms, and so on. [53] Then, scientists redefined AI as *a computer-based system that duplicates the most important human abilities: thinking and finding causes.* [54] Then AI becomes a technology that has two sides; the first side is to improve human welfare, but the second side has the potential to produce various problems for humans—for example, as a supporter of crime.

4.3 METHODOLOGY OF FORENSIC BETWEEN SCIENCE AND TECHNOLOGY

Forensics, like other sciences, or digital forensics, along with its methodology, [55] is an open system which, in its development, shows that something is related to something else. [56] Digital forensics, with its methodological bridges, interacts with other science or technology as an effort to achieve the Sustainable Development Goals (SDGs). Thus, AI as a technology has taken a role in digital forensics due to its respective characteristics, namely:

1. Forensics has, from the start, influenced other scientific fields, or influenced the potential of all other scientific fields as input for its development and strengthening it. Currently, digital forensics, in particular, incorporates AI based on the same basic concepts and the involvement of computer technology to ensure the system is able to serve according to standards. This nature reveals their respective roles. Digital forensics is evolving due to AI, [57] and AI is improvising with digital forensics. [4] This is called *energy import*. Likewise, AI is an input (energy import) for digital forensics.
2. Forensics, or digital forensics, requires a way to complete it (as a processing procedure) efficiently. This method may digitally involve all the capabilities of AI or, from a different perspective, requires intelligent methods to produce better results. [58] AI is a solution for digital forensics.
3. Digital forensics involving AI provides the necessary evidence based on arguments and facts, where forensics organizes the arguments to place appropriate investigation methods on the case, while AI reveals the facts that support the argument. Thus, the attitude of the environment and everything related to the case determines the importance of forensics. [59] AI is as an output of digital forensics.
4. Digital forensics, together with methodology, with the strengthening of AI, ensures the success of each activity, where AI directly plays the activities of entering, processing, and determining results as a series of activities that take place continuously to guarantee the results of one series of activities as validation, while one other series of activities for the comparison case, and

so on. [57] AI is as a series of events—input, solution, output—for digital forensics.

5. Digital forensics treats AI as negative entropy, where both postulates and methodology require changes to answer the challenges of the times. The challenge states everything undergoes degradation (entropy), so negative entropy reverses that. AI predicts not only results but also the potential (a review) for novelty, that is, the development of digital forensics. [60] AI is as negative entropy for digital forensics.

6. Digital forensics involves all the potential of AI as a result of the input data varying in structure, type, and arrangement, thus producing diverse information in terms of presentation. AI is as information for digital forensics. [61]

7. Digital forensics, with various arguments that require reasoning activities, must be in a stable state where AI plays a role in adapting the results of that reasoning to the demands of providing facts. [62] AI is as a steady state for digital forensics.

8. Digital forensics and AI have different roles. In general, digital forensics has an argument for challenging the truth of existing reality, while AI proves the truth in the right ways. [63] AI is as a differentiation for digital forensics.

9. Integrated digital forensics and AI-developed mechanisms integrate parts that have been specialized based on their respective methods. [64, 65] AI is as coordination for digital forensics.

10. Digital forensics and AI have the same goal (equifinality), which is to reach the truth or evidence that is the target. [66, 67] AI is as equifinality for digital forensics.

In general, forensics recognizes the identity of a forensic object and recognizes the treatment it has experienced so that the object has the potential to lose its identity. In this way, forensics traces the existing reality backward through various methods that allow the presence of facts in the search from the investigation process. Therefore, AI has the potential to play some or all the roles from 1 to 10 according to needs, but actually, the ten roles of AI provide maximum results for digital forensic. [68] Reality provides initial facts as data, as the starting point for recognition, and continuously produces data from new facts that emerge during the investigation. Likewise, in digital forensics, objects have become digital data containing binary numbers which are generally recognized as documents, but all documents definitely have their own identity even though the identity of the forensic object in the document is unknown. [69, 70] Some methods related to recognition are:

1. *Chromatography.* [71] A general technique used to separate mixed components from the mobile phase in forensics. AI provides an overview of separating mixed data by classifying it into various classes according to the characteristics given through labeling. [72]

2. *Toxicology.* [73] A method for detecting poison. AI plays a role in identifying color changes based on chemical and medical process clues. Apart from that, AI has the potential to recognize poisoning through patterns that appear on forensic objects based on existing data, either through laboratory experiments or through previous cases. [74]

3. *Ballistics*. [75] A way of analyzing the pattern formed, usually between the bullet, casing, and barrel, which determines the identity of the tool. In determining the appropriate pattern and sound, AI has a role in making it easier to obtain certainty. [76, 77]

4. *Anthropometry*. [78] A method that relies on physical measurements, which, of course, involve numbers and geometry, but also estimates on gender, age, mortality, and voice. [79, 80] AI, as an intelligent tool, apart from recognizing based on physical size, can also definitely determine body posture and has the potential to recognize the shape of the face behind the mask, as well as recognize sound patterns, gait, and so on that are related to physicality, which may be obtained through sensors and Internet of Things (IoT) equipment. [81]

5. *Fingerprints*. [82] A method for identifying someone based on palms and fingerprints, with the assumption that everyone has different fingerprints. AI plays a role in easily recognizing each fingerprint with the patterns found on each finger. [83, 84]

6. *Uhlenhuth test*. [85] A method for differentiating blood according to sepsis. Human blood is different from animal blood in general. [86] In addition, the same or an improved method can forensically detect diseases caused by animals, such as mosquitoes, [87] for example. AI has the potential to guide recognition through the concept of decay or time elapsed by forensic objects. If it is integrated with data based on a timeline or based on the emergence of several outbreaks and their origins, [88] such as COVID-19, it is possible to reveal plans for the genocide of one nation by another nation based on AI. [89]

7. *DNA*. [90] A method for distinguishing one individual from another individual, identifying them through variations in the genetic sequence. Genetic tracing at the scene of the incident provides information on the identity of the victim and perpetrator. [91] AI plays a role in facilitating identification based on available DNA databases, or recognizing the genetic sequence from the records of criminals. [59]

8. *Maturation*. [92, 93] A method for drawing conclusions from a series of events, which involves multiple characteristics, starting from place and time, tools used, and targets taken based on digital data. [94, 95] In this case, AI, through records, makes it possible to study and predict possibilities related to all these characteristics to determine the characteristics of the next event. [96, 97]

The variety of forensic objects has increased, along with the emergence of computer technology, the presence of networks, and the rise of the internet, but the sophistication of forensic objects has also increased, with the intricacies of protecting crimes by the manipulation of various actual facts. That way, methods must also be improved by adopting various technologies, especially AI technology. [98] In general, manipulation is carried out to kill rival characters, by building slander, changing true information, and sowing hoaxes. The integration of digital forensics and AI has various potentials to reveal the true nature of a forensic object. [99]

4.4   THE WORLD OF DIGITAL AND FORENSICS

In the digital world, according to computer science, computers have the privilege of having a congenital defect known as arithmetic weakness. [100] This weakness always automatically generates errors in computing if the programmers or system designers are not careful or lack the ability. These weaknesses also carry over into software, operating systems, and programming systems. Depending on the machine and software used, each computation is only capable of generating numbers no more than 2^8, 2^{16}, 2^{32}, and so on. [101] This weakness presents the proposition that the arithmetic capability of the machine must be greater than or equal to the arithmetic capability of the software in carrying out the computation. That way, in terms of software engineering or creating any computing system that serves the public and community interests, it is necessary to involve digital forensics (computer) experts or digital auditors. [102] Ethically, it is mandatory that the system development committee involve digital forensics experts or consultants in related fields, to avoid misadministration. [103]

Likewise, arithmetic weaknesses often creep into the system unwittingly or are deliberately introduced to create fraud in calculations. [104] Either intentionally or unknowingly, there are gaps in legal violations, and digital forensic experts must be able to uncover them. Errors in expert consultants or system builders, or both at the same time, or errors involving the entire committee, including those who ordered them. [105] Some considerations related to the weaknesses of arithmetic in the digital world are as follows.

4.4.1   DATA

When it is stated that the smallest unit that has meaning is data, then the data is not only in the form of numbers but also various types that can be recorded in storage. [106] However, today's data is mostly unstructured or semi-structured. Collections of words naturally come from different sources according to their level of validity and become information scattered throughout the world of communication in various forms and arrangements. A collection of words is sometimes intertwined with each other or face each other as sources and comments, or support each other as references in scientific works. Thus, data or information contained on the internet, as documents or the web, has a level of truth, although quite a bit of information from the academic world does not contain errors, but the highest level of information that can be trusted comes from the academic world. [107]

Currently, most governments utilize the digital world to maintain their position. Governments that are unable to manage the country tend to cover up their incompetence by maintaining influencers as buzzers. Moreover, the internet system is controlled by the government specifically for that country. As a result, a lot of information and data are fake (slander) on the internet, and the scientific world must be able to carry out forensics on it if it raises an issue as research material. Not a few of the scientific worlds are polluted by incorrect information (hoaxes); even though the correct procedures are used in processing them, they produce data whose truth cannot be justified. The role of AI is required to classify or group such information through data extraction or mining using correct methods, not a justification. One way is learning where the identities of hoax generators are entered into a classification or dataset, which functions to reduce their influence in cyberspace. [108] Therefore, if it is not reduced, people who have no performance will emerge as people who have achievements, for example.

The quality of work is not just a number but a legitimate achievement of the activity. The quality of activities comes from before the work (intention and planning) or events that precede it, during the time the work is carried out, and after the work is carried out. AI integrated into digital forensics tracks all issues to see the relationship between one another based on the identities of data or information and the flow of information originating from interrelated activities. [109]

When the number is related to humans—for example, the number of humans—it cannot be said to be just a number, because it violates human rights, because eliminating a human child means eliminating the entire human being. Numbers are an absolute consideration; in this case, numbers are not just statistics, but numbers provide protection for human dignity. Therefore, in the world of computing, where the input is a number, it is necessary to consider whether the number is of the numeric type or of the character (string) type. When a number of the numeric type is confronted with any operator, it produces the appropriate number value, for example, $1 + 2 + 3 = 6$, if the numbers 1, 2, 3 and the operator $+$ are of the numeric type. On the other hand, when any number is seen as a character symbol, for example, $1 + 2 + 3 = 123$ (always expressed as $'1' + '2' + '3' = '123'$) if 1, 2, and 3 are symbols of numbers and the $+$ operator is a string operator. Based on that, if there is a limit set in numbers, for example, not more than or equal to 300 and not smaller than 0 (null), then the number accepted is one of $1, 2, 3, \ldots, 300$. On the other hand, if the number symbol is a reference to the type of number entered, potential data of more than 300 or minus -44 can be accepted, although it may be stipulated that the number of characters allowed is three digits. [110] Digital forensics simply looks at the potential for errors, but searching for quite a lot of data involves computing, which is often assisted by AI in providing conclusions about the behavior of the computing environment. [60]

4.4.2 COMPUTING SYSTEMS

Computing systems include all information systems or applications that either directly or indirectly involve computing in that system. [111] Computing systems usually involve consumers or individuals who socially interact with the computing system, where organizations, companies, or governments provide services in the interest of individuals based on the system by recording their respective data into the system. In this way, the computing system is directly responsible for protecting each individual's personal data, and the system owner provides services fairly and transparently for individuals. [112] Any unfairness and non-transparency may challenge system services and may be considered a violation or a crime. If based on digital forensics, it violates applicable regulations. After all, every transaction in a computing system has a track record, and certain violations can be traced by any AI search method. [113]

Each computing system provides a form as a user interface with a database, and the form contains information that is entered temporarily before being recorded into the database by the system. Between the forms, fill in different types of data according to the data requested. The numeric data type requires a numeric form to be entered, while others include characters or strings, dates, and others. If the form is of a different type to the type of data intended, such as a numeric number having a form of character type, then there are two possibilities that apply: an error occurred when designing the system, and there is also the potential for an error in the database, making it possible to enter a number that exceeds the allowable number capacity. The second is due to accommodating very large numbers through a string-type form, but in computing, it involves computing modulo numbers so that the computational results are in accordance with the aims and objectives. Likewise, regardless of the intended use, there are three things related to numeric data: the entered numeric value, the input form, and the database metadata, which should be in sync with each other. Regarding the type of numeric data, such as monetary value or entity value, which differ from one another, financial data synchronization has an input form of real number numeric type, and database metadata is of real number numeric type as well, while entity data synchronization is always integer type for those three things. [114] Unsynchronization testing is carried out by creating dummy data that is entered into the system before launching the system, then performing any computation, then discarding it before using the computing system. Or by conducting digital forensics, comparing design documents with the interface and structure of computing systems and databases. The compared facts can be recognized by AI to obtain appropriate suitability values for a computing system to properly serve the interests of many people. For example, relating to confirmations received by users regarding transactions they carry out. [115]

Because computing systems are so complex to build, they require a group of institutionally experienced experts, and system development takes place when a contract is in place. A computing system has design, planning, and implementation, then it is launched in a place that also has its own security. Because the computing system serves many people from various places, it is placed where security is guaranteed—especially, the database must be accessible from various places or be in cloud computing space, where the database has backups in different places with conditions that are more secure than natural disasters and protected from human invasion. [116] In contrast to that, the computing system used by the government to serve the public must be located in the country's territory, in accordance with applicable laws that state that the government is obliged to manage land, water, air, and everything in it for the benefit of the nation and state. [117] Therefore, the computing system used by the government is not suitable for being outside the country, and it is considered to violate the nation's independence as an independent country. [118] Forensically, contracts that are not in accordance with a country's laws have the potential to violate regulations, and tracing errors in contract procurement can be done from the digital record between the person who received the contract and the person who gave the contract. By involving AI, digital forensics based on the track record states some conclusion about error: whether it is a procedural error, systematic error, abuse of authority, or ultimately, a betrayal of the state. [119, 120]

4.4.3   SOCIAL NETWORKS AND SOCIAL MEDIA

Social media is a place to collect various data in terms of size, relationship, type, shape, structure, dimensions, source (origin), knowledge, dimensions, aims, and objectives. Social media generally acts as a data platform where information after information is collected into big data. [121–123] Therefore, the use of social media as a source of information has the potential to be misleading. Information that comes from processing social media content without filtering, especially not involving AI to do forensics, provides conclusions that are not suitable for most people according to common sense. Social media describes the social interactions of social members and the activities of those who collectively hold social discussions to increase their closeness, but it cannot be avoided that social media contains the dark side of governments in the world. [124, 125]

It is a law of nature that people with similar interests tend to be in one group according to social networks. In many incidents, criminals, corruptors, traitors, and fraudsters are in their communities. Although not more closed, academics are also in their communities. [126, 127] One with those in the community who usually support each other and when they support others outside their community or they are supported by others from outside their community. Indirectly, this requires digital forensics to search for traces of behavior of outsiders through various available information streams to see whether they participated in a crime, [128] and to facilitate the search, it involves AI as a tool to learn behavior by behavior. [129]

4.4.4   INDIVIDUAL BEHAVIOR

Initially, it was difficult to predict human behavior based on previous events. [130] However, information recorded in digital space makes it possible to explore it in order to search for an individual's potential to become a psychologically good or bad human being. [131] The aim is to ensure social security. The attitudes and habits of each individual may have been accidentally recorded from previous events, and individuals are too ambitious in certain matters, so they carry out manipulation, such as forging documents: fake diplomas, falsified hereditary history, and so on. [132–135]

Initially, forensics as a science made it possible to reveal lies through attitudes, mimics, or body responses to questions through lie tests, but through the flow of information about individuals in the digital space, not only lies, but also their behavior, can potentially be predicted and confirmed by AI. [136, 137]

4.4.5   COMPLEXITY AND DIMENSIONALITY

Information in an information space, data in any database, or a polyglot has its own complexity due to its dimensions. Dimensions come from the features that characterize an entity. Human entities, for example, consist of ethnicity, religion, education, region, wealth, modesty or not, and so on, which make it impossible to hastily summarize them into smaller dimensions. [138] Human behavior and attitudes towards something always differ based on all those attributes. That way, sampling is not easy to do so that it is easy to get information. The sample may be a description of the population, as long as it complies with sampling procedures. Sampling reduces not only the amount of data but also the dimension of the data, and usually the data behaves randomly [139]—to perpetuate fast calculations, or may be as quick count, for example. In developed countries, such as the United States, the results of democratic general elections are known after several days; in Indonesia, the results are known before the election day. Whether it is a joke or a real incident depends on how it is revealed as a forensic object. [140, 141]

Sampling requires mapping data and dimensions and projecting them into a concise picture to obtain an appropriate sample to represent the population. Projections are carried out by AI as forensics for errors in sampling. That way, AI concludes whether the fast calculation is a trick or not, based on the forensics carried out on the sample. [142]

4.5 CONCLUSION: TOWARD FORENSIC INTELLIGENCE

One of the roles of artificial intelligence is to enable the development of forensic science as a downstream process. [143, 144] The combination of digital forensic science and artificial intelligence technology improves forensic performance in identifying objects and the treatment received by the objects. Identification is carried out by exploring, recovering, investigating, recording, and analyzing the current condition of the object so as to reveal at least the identity of the object, or to trace the treatment that has been applied to the object. Digital recording has the principle of 0s and 1s in digital forensics to build documentation or to record it as data or implement the results into a control system that integrates all forensic activities to provide lessons about each event behavior and its conclusions, if necessary. [145] In this way, digital forensics acts as an open system that requires the import of energy (input by input), the completion of the process, results/outcomes, the possibility of repeated events, negative entropy, and information and its additions, always in a steady state. There are differences that provide uniqueness (perhaps identity), integrity, or coordination and equipped with the principle of one goal. [146, 147]

On the other hand, AI basically recognizes an object more deeply based on its characteristics and data, eliminates or reduces unimportant things such as dimensions and structure, analyzes differences and similarities, examines and studies available and related records and data, and sorts parts by parts—part of an object according to data instructions or labels. In addition, AI has the potential to collect appropriate data/information into its structure, predict the increasing form of events, and build a reasoned conclusion. [148]

AI in general is a digital forensic tool that is able to adapt to all fields of reality. AI contains methods that have performance in accordance with forensic issues, methods that are able to analyze objects, such as differences between two images that are similar but not the same, detect fraud in transactions, and others. So the target of forensics and AI, apart from objects, is the treatment of objects that are tangible and intangible, where AI makes it possible to simulate images before and after an event. [149, 150] In the digital world, the object may be data, storage, database, service system, information flow, or directly, the technology itself, such as computers, computer networks, electronic equipment, and so on—or everything related to the equipment, such as installation, position, and location, which reveal log-in and intended use. Thus, the potential of AI in digital forensics is in the definition of *forensic intelligence* [151, 152]:

> *A science that aims to intelligently recognize—explore, recover, investigate, record, and analyze the conditions of objects—and express the impact of natural touch, organisms, or anything man-made on objects or entities based on the principle of applying 0s and 1s, by completing information extraction or data mining.*

BIBLIOGRAPHY

1. Schelenz, L. (2023). Diversity concepts in computer science and technology development: A critique. *Science Technology and Human Values*, *48*(5), 1054–1079. DOI 10.1177/01622439221122549
2. Dragonas, E., Lambrinoudakis, C., & Kotsis, M. (2023). IoT forensics: Exploiting unexplored log records from the HIKVISION file system. *Journal of Forensics Sciences*, *68*(6), 2002–2011. DOI 10.1111/1556-4029.15349
3. Saha, B., Khan, A. K., Lalitha, V. L., & Prasad, L. V. N. (2023). A digital forensics model for the examination of QR code and android app to investigate Aadhaar card identity fraud. *Journal of Applied Security Research*. DOI 10.1080/19361610.2023.2279399
4. Al-Mugern, R., Othman, S. H., & Al-Dhaqm, A. (2024). An improved machine learning method by applying cloud forensic meat-model to enhance the data collection process in cloud environments. *Engineering, Technology and Applied Science Research*, *14*(1), 13017–13025. DOI 10.48084/etasr.6609
5. Farouk, S., Osman, A. M., Awadallah, S. M., & Abdelrahman, A. S (2023). The added value of using artificial intelligence in adult chest X-rays for nodules and masses detection in daily radiology practice. *Egyptian Journal of Radiology and Nuclear Medicine*, *54*(1). DOI 10.1186/s43055-023-01093-y
6. Lim, T. (2024). Environmental, social, and governance (ESG) and artificial intelligence in finance: State-of-the-art and research takeaways. *Artificial Intelligence Review*, *57*(4). DOI 10.1007/s10462-024-10708-3
7. Shenoy, R. P. (1985). Artificial intelligence—The emerging technology. *Defence Science Journal*, *15*(2), 135–149. DOI 10.14429/dsj.35.6004
8. Horsman, G. (2023). Digital evidence strategies for digital forensic science examinations. *Science and Justice*, *63*(1), 116–126. DOI 10.1016/j.scijus.2022.11.004
9. Alderson. (1828). Forensic cant. *The Lancet*, *10*(265), 812–813. DOI 10.1016/S0140-6736(02)98124-9
10. Kasprzak, J., & Kasprzak, M. (2021). Forensics as a science and practice. *Studia Iuridica Lublinensia*, *30*(4), 315–331. DOI 10.17951/sil.2021.30.4.315-331
11. Jota Baptista, C., Seixas, F., Gonzalo-Orden, J. M., & Oliveira, P. A. (2022). Wildlife forensic sciences: A tool to nature conservation towards a one health approach. *Forensic Sciences*, *2*(4), 808–817. DOI 10.3390/forensicsci2040058
12. Svendsen, B. B. (1977). Is the existence of a forensic psychiatry justified? On forensic psychiatry in the scandinavian countries. *Acta Psychiatrica Sandinavica*, *55*(3), 161–164. DOI 10.1111/j.1600-0447.1977.tb00154.x
13. Shubik, M. (1960). Bibliography on simulation, gaming, artificial intelligence and allied topics. *Journal of the American Statistical Association*, *55*(292), 736–751. DOI 10.1080/01621459.1960.10483374
14. Hsu, T.-C., Chang, C., & Jen, T.-H. (2023). Artificial intelligence image recognition using self-regulation learning strategies: Effects on vocabulary acquisition, learning anxiety, and learning behaviours of English language learners. *Interactive Learning Environments*. DOI 10.1080/10494820.2023.2165508

15. Birenbaum, M. (2021). The Chatbots' challenge to education: Disruption or destruction? *Education Sciences*, *13*(7). DOI 10.3390/educsci13070711

16. Nasution, M. K. M., Syah, B. R., & Elveny, M. (2023). What is data science. *Data Science with Semantic Technologies*. DOI 10.1201/9781003310785-1

17. Iorliam, A. (2018). History of forensic science. *Springer Briefs in Computer Science*, 3–16. DOI 10.1007/978-3-319-94499-9_2

18. Koppl, R. (2007). Diversity and forensics: Diversity in hiring is not enough. *Medicine, Science and the Law*, *42*(2), 117–124. DOI 10.1258/rsmmsl.47.2.117

19. Kip, H., Kelders, S. M., Bouman, Y. H. A., & Van Gemert-Pijnen, L. J. E. W. C. (2019). The importance of systematically reporting and reflecting on eHealth development: Participatory development process of a virtual reality application for forensic mental health care. *Journal of Medical Internet Research*, *21*(8). DOI 10.2196/12972

20. Erzincllio˜glu, Y. Z. (1983). The application of Entomology to forensic medicine. *Medicine, Science and the Law*, *21*(1), 57–63. DOI 10.1177/002580248302300110

21. Smith, J. G. (1829). Forensic medicine. Case of the late Mr. Neale and Butler, the soldier, medico-legally considered. *The Lancet*, *12*(294), 84–88. DOI 10.1016/S0140-6736(02)92328-7

22. Caggiani, M. C., & Colomban, P. (2024). Advanced procedures in Raman forensic, natural, and cultural heritage studies: Mobile set-up, optics, and data treatment—state of the art and perspectives. *Journal of Roman Spectroscopy*, *55*(2), 116–124. DOI 10.1002/jrs.6633

23. Tan, L., & Byard, R. W. (2024). Cardiac amyloid deposition and the forensic autopsy—a review and analysis. *Journal of Forensic and Legal Medicine*, *103*. DOI 10.1016/j.jflm.2024.102663

24. Ruder, T. D., Kuhnen, S. C., Zech, W.-D., Klaus, J. B., Lombargo, P., & Ith, M. (2023). Standards of practice in forensic age estimation with CT of the medical clavicular epipshysis—a systematic review. *International Journal of Legal Medicine*, *137*(6), 1757–1766. DOI 10.1007/s00414-023-03061-7

25. Morgan, R. M., & Bull, P. A. (2007). The philosophy, nature and practice of forensic sediment analysis. *Progress in Physical Geography*, *31*(1), 43–58. DOI 10.1177/0309133307073881

26. McLeod, D. A., Natale, A. P., & Mapson, K. W. (2023). *Handbook of Forensic Social Work: Theory, Policy, and Fields of Practice*. Oxford University Press. DOI 10.1093/oso/9780197694732.001.0001

27. Klotzbach, H., Krettek, R., Bratzke, H., & Pu¨schel, K. (2004). The history of forensic entomology in German-speaking countries. *Forensic Science International*, *144*(2–3), 259–263. DOI 10.1016/j.forsciint.2004.04.062

28. Hauptman, L. M. (2010). Beyond forensic history: Observations based on a forty-year journey through iroquois country. *Journal of the West*, *49*(4), 11–19.

29. Parisis, N. (1990). Toxicological analysis of antimony in human biological materials. Application in forensic science. *Journal of Pharmacie de Belgique*, *45*(5), 334.

30. Bertsatos, A., Garoufi, N., Koliaraki, M., & Chovalopoulou, M.-E. (2023). Paving new ways in forensic contexts with virtual osteology applications: CSG-toolkit—Dosteology package for cross-sectional geometry analysis. *Annals of 3D Printed Medicine*, *9*. DOI 10.1016/j.stlm.2022.100094

31. Olkhovsky, V., Gubin, M., Grygorian, E., Khyzhniak, V., & Knigavko, O. (2021). Actual problems in forensic medical assessment of trauma of the respiratory organs. *Journal of Punjab Academy of Forensic Medicine and Toxicology*, *21*(2), 27–32. DOI 10.5958/0974-083X.2021.00055.8

32. Gelir, I. (2022). Preschool children learn physics, biology, chemistry and forensic science knowledge with integrated teaching approaches. *International Journal of Early Years Education*. DOI 10.1080/09669760.2022.2037077

33. Varderkolk, J. (2002). Forensic science, physhology and philosophy. *Journal of Forensic Identification, 52*(3), 252–253.

34. Panigrahi, A. (2019). Liquid biopsy: A potential tool in oral forensic science. *Indian Journal of Forensic Medicine and Toxicology, 13*(4), 1936–1939. DOI 10.5958/0973-9130.2019.00604.2

35. Mandal, A., & Amilan, S. (2023). Evaluating the perceived usefulness and fairness of forensic accounting and investigation standards. *Journal of Financial Regulation and Compliance, 11*(5), 754–769. DOI 10.1108/JFRC-12-2022-0157

36. Blackwell, R. J., & Crisci, W. A. (1975). Digital image processing technology and its application in forensic sciences. *Journal of Forensic Sciences, 20*(2), 288–304. DOI 10.1520/jfs10276j

37. Palmer, G. (2001). A road map for digital forensic research. *Proceedings of the Digital Forensic Research Conference, DFRWS 2001 USA,* iii–42. Digital Forensic Research Workshop.

38. Rogers, M., Losavio, M., Wilson, D., Elder, D., Elmaghraby, A., & Srinivasan, S. (2006). Issues in building the digital forensics bridge from computer science to judicial science. *6th Annual Digital Forensic Research Workshop, DFRWS 2006.* http:/dfrws.org.

39. Persin, S. M. (1964). Information content in digital measurements. *Measurement Techniques, 7*(7), 558–563. DOI 10.1007/BF00980030

40. Chu, H.-C., Deng, D.-J., & Chao, H.-C. (2011). An ontology-driven model for digital forensics investigations of computer incidents under the ubiquitous computing environments. *Wireless Personal Communications, 56*(1), 5–19. DOI 10.1007/s11277-009-9886-x

41. Welch, T. (1977). Computer crime investigation and computer forensics. *Information Systems Security, 6*(2), 56–80. DOI 10.1080/10658989709342536

42. Tessarolo, A. A., & Marignani, A. (1996). Forensic science and the internet. *Journal of the Canadian Society of Forensic Science, 29*(2), 87–92. DOI 10.1080/00085030.1996.10757051

43. Nasution, M. K. M. (2019). Forensic in information technology: A redefinition. *Journal of Physics: Conference Series, 1235*(1). DOI 10.1088/1742-6596/1235/1/012106

44. Reeves, S. (2013). Human-computer interaction issues in human computation. *Handbook of Human Computation,* 411–419. Springer. DOI 10.1007/978-1-4614-8806-4_32

45. Kampis, L. (2016). Garden of Eden for artificial intelligence: How "the Talos principle" demonstrates the difficulty of defining consciousness for AI on the implied player. *AISB Annual Convention 2016, AISB 2016.* The Society for the Study of Artificial Intelligence and the Simulation of Behaviour (AISB).

46. Hashim, F. H., & Abdullah, W. N. W. (2017). Swarm intelligence: From the perspective of Al-quran and Al-sunnah to natural and artificial systems. *Advanced Science Letters, 23*(5), 4580–4585. DOI 10.1166/asl.2017.8996

47. Gielen, S. (2007). Natural intelligence and artificial intelligence: Bridging the gap between neurons and neuro-imaging to inderstand intelligent behaviour. *Studies in Computational Intelligence, 63,* 145–161. DOI 10.1007/978-3-540-71984-7_7

48. Haenlein, M., & Kaplan, A. (2019). A brief history of artificial intelligence: On the past, present, and future of artificial intelligence. *California Management Review, 51*(4), 5–14. DOI 10.1177/0008125619864925

49. Buchanan, B. G. (2005). A (very) brief history of artificial intelligence. *Al Magazine, 26*(4), 53–60.

50. Frank, P. M. (2004). Chess playing machines from natural towards artificial intelligence? *Proceedings of the International Conference on Systems Science, 1,* 168–181.

51. Wiegand, T., Krishnamurthy, R., Kuglitsch, M., Lee, N., Pujari, S., Salath´e, M., Wenzel, M., & Xu, S. (2019). WHO and ITU establish benchmarking process for artificial intelligence in health. *The Lancet, 394*(10192), 9–11. DOI 10.1016/S0140-6736(19)30762-7

52. Jaurez, J. J., & Radhakrishnan, B. D. (2022). Application of artificial intelligence and the Cynefin framework to establish a statistical system prediction and control (SSPC) in engineering education. *ASEE Annual Conference and Exposition, Conference Proceedings*. American Society for Engineering Education.

53. Wenting, Z., Yuanbiao, Z., Weixia, L., Zhongjle, S., Wenqi, Z., & Yiming, P. (2012). River trip optimization scheduling based on artificial intelligence simulation and the bee-swarm genetic algorithm. *Research Journal of Applied Sciences, Engineering and Technology*, *4*(19), 3801–3810.

54. Saad, A., Saad, M., Emaduddin, S. M., & Ullah, R. (2020). Optimization of bug reporting system (BRS): Artificial intelligence based method to handle duplicate bug report. *Communication in Computer and Information Science*, *1198*, 118–128. DOI 10.1007/978-981-15-5232-8_1155.

55. Nasution, M. K. M. (2020). Methodology. *Journal of Physics: Conference Series*, *1566*(1). DOI 10.1088/1742-6596/1566/1/012031

56. Singh, S., & Singh, V. K. (2023). Digital forensic investigation: Ontology, methodology, and technological advancement. *Advancement in Cybercrime Investigation and Digital Forensics*, 137–160. DOI 10.1201/9781003369479-7

57. Dunsin, D., Ghanem, M. C., Ouazzane, K., & Vassilev, V. (2024). A comprehensive analysis of the role of artificial intelligence and machine learning in modern digital forensics and incident response. *Forensic Science International: Digital Investigation*, *48*. DOI 10.1016/j.fsidi.2023.301675

58. Gupta, R., Mane, M., Bhardwaj, S., Nandekar, U., Afaq, A., Dhabliya, D., & Pandey, B. K. (2023). Use of artificial intelligence for image processing to aid digital forensics: Legislative challenges. *Handbook of Research on Thrust Technologies? Effect on Image Processing*, 433–447. IGI Global. DOI 10.4018/978-1-6684-8618-4.ch026

59. Singh, K., Poongodi, T., & Sagar, S. (2021). Various applications of artificial intelligence in digital forensic investigation. *Indian Journal of Forensic Medicine and Pathology*, *14*(Special Issue 2), 263–269. DOI 10.21088/ijfmp.0974.3383.14221.36

60. Verma, R., Garg, S., Kumar, K., Gupta, G., Salehi, W., Pareek, P. K., & Knieˇzova, J. (2023). New approach of artificial intelligence in digital forensic investigation: A literature review. *Lecture Notes in Networks and Systems (LNNS)*, *735*, 399–409. DOI 10.1007/978-3-031-37164-6_30

61. Du, X., Hargreaves, C., Sheppard, J., Anda, F., Sayakkara, A., Le-Khac, N.-A., & Scanlon, M. (2020). SoK: Exploring the state of the art and the future potential of artificial intelligence in digital forensic investigation. *ACM International Conference Proceedings Series*. Association for Computing Machinery. DOI 10.1145/3407023.3407068

62. Mitchell, F. R. (2014). An overview of artificial intelligence based pattern matching in a security and digital forensic context. *Cyberpatterns: Unifying Design Patterns with Security and Attack Patterns*, 215–222. DOI 10.1007/978-3-319-04447-7_17; ISBN: 9783319044477

63. Manasa, S., & Kumar, K. P. (2022). Digital forensic investigation for attacks on artificial intelligence. *ECS Transactions*, *107*(1), 19639–19645. DOI 10.1149/10701.19639ecst

64. Costantini, S., de Gasperis, G., & Olivieri, R. (2019). Digital forensics and investigations meet artificial intelligence. *Electronic Proceedings in Theoretical Computer Science*, *EPTCS*, 306. Open Publishing Association.

65. Costantini, S., de Gasperis, G., & Olivieri, R. (2019). Digital forensics and investigations meet artificial intelligence. *Annals of Mathematics and Artificial Intelligence*, *86*(1–3), 193–229. DOI 10.1007/s10472-019-09632-y

66. Punjabi, S. K., & Chaure, S. (2022). Forensic intelligence—combining artificial intelligence with digital forensics. *2022 2nd International Conference on Intelligent Technologies, CONIT 2022*. Institute of Electrical and Electronics Engineers Inc. DOI 10.1109/CONIT55038.2022.9848406

67. Reedy, P. (2022). Artificial intelligence in digital forensics. *Encyclopedia of Forensic Sciences*, Volumes 1–4, Third Edition, 170–192. Elsevier. DOI 10.1016/B978-0-12-823677-2.00236-1

68. Hall, S. W., Sakzad, A., & Minagar, S. (2022). A proof of concept implementation of explainable artificial intelligence (XAI) in digital forensics. *Lecture Notes in Computer Science (Including Subseries Lecture Notes in Artificial Intelligence and Lecture Notes in Bioinformatics), 13787*, 66–85. DOI 10.1007/978-3-031-23020-2_4

69. Amjed, A., Mahmood, B., & Almukhtar, K. A. K. (2022). Approaches for forgery detection of documents in digital forensics: A review. *Communications in Computer and Information Science (CCIS), 1548*, 335–351. DOI 10.1007/978-3-030-97255-4_25

70. Shankar, G., Al-Farhani, L. H., Anitha Christy Angelin, P., Singh, P., Alqahtani, A., Singh, A., Kaur, G., & Samori, I. A. (2023). Improved multisignature scheme for authenticity of digital document in digital forensics using Edward-curve digital signature algorithm. *Security and Communication Networks, 2023*. DOI 10.1155/2023/2093407

71. Palmieri, V. M., & Romano, C. (1953). Paper partition chromatography in forensic medicine. *Acta medicinae legalls et socialis*, 6(3–4), 239–243.

72. Faisal, K., Olatunji, S. O., & Ghouti, L. (2009). Classification of premium and regular gasoline using support vector machines as novel approach for arson and fuel spill investigation. *Proceedings of the 2009 International Conference on Artificial Intelligence, ICAI 2009, 1*, 345–350. United States Military Academy, Network Science CenterHST Harvard University MIT, Biomedical Cybernetics Lab.

73. Spiehler, V. R. (1989). Computer-assisted interpretation in forensic toxicology: Morphine-involved deaths. *Journal of Forensic Sciences*, 34(5), 1104–1115. DOI 10.1520/jfs12747j

74. Cechner, R. L., & Sutheimer, C. A. (1990). Automated rule-based decision systems in forensic toxicology using expert knowledge: Basic principles and practical applications. *Journal of Analytical Toxicology*, 14(5), 280–284. DOI 10.1093/jat/14.5.280

75. Petty, C. S., & Hauser, J. E. (1968). Rifled shotgun slugs, wounding and forensic ballistics. *Journal of Forensic Sciences*, 13(1), 114–123.

76. Raponi, S., Oligeri, G. & Ali, I. M. (2022). Sound of guns: Digital forensics of gun audio samples meets artificial intelligence. *Multimedia Tools and Application*, 81(21). DOI 10.1007/s11042-022-12612-w

77. Piraianu, A.-I., Fulga, A., Musat, C. L., Clobotaru, O.-R., Poalelungi, D. G., Stamate, E., Ciobotaru, O., & Fulga, I. (2023). Enhancing the evidence with algorithms: How artificial intelligence is transforming forensic medicine. *Diagnostics*, 13(18). DOI 10.3390/diagnostics13182992

78. Pola, F. M., S,tef˘anescu, C. L., & Corici, P.-D. (2009). Forensic value of mandibular anthropometry in gender and age estimation. *Romanian Journal of Legal Medicine*, 17(1), 45–50. DOI 10.4323/rjlm.2009.45

79. Davy-Jow, S. L., Lees, D. M. B., & Russell, S. (2013). Virtual forensic anthropology: Novel applications of anthropometry and technology in a child death case. *Forensic Science International*, 224(1–3), e7. DOI 10.1016/j.forsciint.2012.11.002

80. Singh, R., Raj, B., & Gencaga, D. (2016). Forensic anthropometry from voice: An articulatory-phonetic approach. *2016 39th International Convention on Information and Communication Technology, Electronics and Microelectronic, MIPRO 2016—Proceedings*, 1375–1380. Institute of Electrical and Electronics Engineers Inc. DOI 10.1109/MIPRO.2016.7522354

81. Grojek, A. E., & Sikos, L. F. (2022). Ontology-driven artificial intelligence in IoT forensics. *Breakthroughs in Digital Biometrics and Forensics*, 257–286. Springer International Publishing. DOI 10.1007/978-3-031-10706-1_12

82. Ikuesan, A. R., & Venter, H. S. (2019). Digital behavioral-fingerprint for user attribution in digital forensics: Are we there yet? *Digital Investigation*, 30, 73–89. DOI 10.1016/j.diin.2019.07.003

83. Horsman, G., Page, H., & Beveridge, P. (2018). A preliminary assessment of latent fingerprint evidence damage on mobile device screens caused by digital forensic extractions. *Digital Investigation, 27*, 47–56. DOI 10.1016/j.diin.2018.10.002

84. Sheldon, A. (2005). The future of forensic computing. *Digital Investigation, 2*(1), 31–35. DOI 10.1016/j.diin.2005.01.005

85. Rice, J. B., & Barber, M. A. (1935). Malaria studies in Greece. A modification of the Uhlenhuth-Weidanz precipitin test for determining the source of blood meals in mosquitoes and other insects. *The Journal of Laboratory and Clinical Medicine, 20*(8), 876–883.

86. Hopkins, M. L. (1935). Development of the thyroid gland in the chick embryo. *Journal of Morphology, 58*(2), 585–613. DOI 10.1002/jmor.1050580211

87. Edrissian, G. H., & Hafizi, A. (1982). Application of enzyme-linked immunosorbent assay (ELISA) to identification of Anopheles mosquito bloodmeals. *Transactions of the Royal Society of Tropical Medicine and Hygiene, 76*(1), 54–56. DOI 10.1016/0035-9203(82)90017-7

88. Nasution, M. K. M. (2022). World on data perspective. *World, 3*(3), 736–752. DOI 10.3390/world3030041

89. Hamburg, D. A., & Wiesel, E. (2015). *Preventing Genocide: Practical Steps toward Early Detection and Effective Action*. Taylor and Francis. DOI 10.4324/9781315632599

90. Dodd, B. E. (1985). Forensic science: DNA fingerprinting in matters of family and crime. *Nature, 318*(6046), 506–507. DOI 10.1038/318506a0

91. Nogel, M. (2024). Some area where digital forensics can support the addressing of legal challenges linked to forensic genetic genealogy. *Forensic Science International Digital Investigation, 49*. DOI 10.1016/j.fsidi.2024.301696

92. Schmeling, A., Reisinger, W., Loreck, D., Vendura, K., Markus, W., & Geserick, G. (2000). Effects of ethnicity on skeletal maturation: Consequences for forensic age estimations. *International Journal of Legal Medicine, 113*(5), 253–258. DOI 10.1007/s004149900102

93. Klepinger, L. L. (2001). Stature, maturation variation and secular trends in forensic anthropology. *Journal of Forensic Science, 46*(4), 788–790. DOI 10.1520/jfs15048j

94. Duranti, L., & Jansen, A. (2009). Authenticity of digital records: An archival diplomatics framework for digital forensics. *4th International Conference on Information Warfare and Security, ICIW 2009*, 134–139. Academic Conferences Ltd.

95. Casey, E. (2019). Maturation of digital forensics. *Digital Investigation, 29*, A1–A2. DOI 10.1016/j.diin.2019.05.002

96. Alnagar, M. H., Gajendrareddy, P., Lee, M. K., & Allareddy, V. (2023). Artificial intelligence and orthodontic practice: The future unveiled. *Integrated Clinical Orthodontics*, 565–575. DOI 10.1002/9781119870081.ch25

97. Mendes, J. Lima, J., Costa, L. A., Rodrigues, N., Leitˉao, P., & Pereira, A. I. (2024). An artificial intelligence-based method to identify the stage of maturation in olive oil mills. *Communication in Computer and Information Science (CCIS), 1982*, 63–77. DOI 10.1007/978-3-031-53036-4_5

98. Verma, M. (2003). Artificial intelligence applied to computer forensics. *Advancements in Cybercrime Investigation and Digital Forensics*, 301–318. Apple Academic Press. DOI 10.1201/9781003369479-15

99. Tynan, P. (2024). The integration and implications of artificial intelligence in forensic science. *Forensic Science, Medicine, and Pathology*. Springer. DOI 10.1007/s12024-023-00772-6

100. Nasution, M. K. M., Hidayat, R., & Syah, R. (2022). Computer science. *International Journal on Advanced Science, Engineering and Information Technology, 12*(1), 1142–1159. DOI 10.18517/ija-seit.12.3.14832

101. Hemery, M., & Fages, F. (2024). On a model of online analog computation in the cell with absolute functional robustness: Algebraic characterization, function compiler and error control. *Theoretical Computer Science*, *991*. DOI 10.1016/j.tcs.2024.114432

102. Fakiha, B. (2023). The role of raspberry Pi in forensic computer crimes. *Journal of Internet Service and Information Security*, *13*(4), 76–87. DOI 10.58346/JISIS.2023.I4.005

103. Bowers, C. M. (2014). Forensic expert ethics: Cases and concepts about ethical forensic practice and testimony in court. *Forensic Testimony: Science, Law and Expert Evidence*, 207–220. DOI 10.1016/B978-0-12-397005-3.00011

104. Pollett, C., & Pruim, R. (2002). Strengths and weaknesses of LH arithmetic. *Mathematical Logic Quarterly*, *48*(2), 221–243. DOI 10.1002/1521-3870(200202)48:2¡221::AID-MALQ221¿3.0.CO;2-V

105. Melinek, J., Thomas, L. C., Oliver, W. R., Schmunk, G. A., & Weedn, V. W. (2013). National association of medical examiners position paper: Medical examiner, coroner, and forensic pathologist independence. *Academic Forensic Pathology*, *3*(1), 93–98. DOI 10.23907/2013.013

106. Nasution, M. K. M. (2022). Understanding data toward going to data science. *Lecture Notes in Networks and Systems (LNNS)*, *503*, 478–489. DOI 10.1007/978-3-031-09073-8_42

107. Nasution, M. K. M., & Elveny, M. (2022). Data modeling as emerging problems of data science. *Data Science with Semantic Technologies: Theory, Practice and Application*, 71–90. Wiley & Sons. DOI 10.1002/9781119865339.ch3

108. Nasution, M. K. M., & Syah, R. (2022). Data management as emerging problems of data science. *Data Science with Semantic Technologies: Theory, Practice and Application*, 91–104. Wiley & Sons. DOI 10.1002/9781119865339.ch4

109. Nasution, M. K. M., Aulia, I., & Elveny, M. (2019). Data. *Journal of Physics: Conference Series*, *1235*(1). DOI 10.1088/1742-6596/1235/1/012110

110. Du, Y. R., Zhu, L., & Cheng, B. K. L. (2019). Are numbers not trusted in a "post-truth" era? An experiment on the impact of data on new credibility. *Electronic News*, *13*(4), 179–195, 2019. DOI 10.1177/1931243119883839

111. Gracon, T. J., Nolby, R. A., & Sansom, F. J. (1971). A high performance computing system for time critical applications. *AFIPS Conference Proceedings—1971 Fall Joint Computer Conference, AFIPS 1971*, 549–560. Association for Computing Machinery, Inc.

112. Chen, W., Xie, G., Li, R., Bai, Y., Fan, C., & Li, K. (2017). Efficient task scheduling for budget constrained parallel applications on heterogeneous cloud computing systems. *Future Generation Computer Systems*, *74*, 1–11. DOI 10.1016/j.future.2017.03.008

113. Taylor, M., Haggerty, J., Gresty, D., & Lamb, D. (2011). Forensic investigation of cloud computing systems. *Network Security*, *2011*(3), 4–10. DOI 10.1016/S1353-4858(11)70024-1

114. Tani, K. I. (1986). Interfaces for intelligent computing systems. *Soviet Journal of Computer and System Sciences*, *24*(6), 44–58.

115. Sharma, V., Vashishth, T. K., Kumar, B., Sharma, K. K., Panwar, R., & Chaudhary, S. (2023). Brains-computer interface: Bridging the gap between human brain and computing system. *2023 IEEE International Conference on Research Methodologies in Knowledge Management, Artificial Intelligence and Telecommunication Engineering, RMKMATE 2023*. Institute of Electrical and Electronics Engineers Inc. DOI 10.1109/RMK-MATE59243.2023.10369702

116. Abrosimov, L. I. (1983). Computation of characteristics of computing systems of complex configuration by means of contours. *Engineering Cybernetics*, *21*(5), 79–86.

117. Jain, R., & Wullert II, J. (2002). Challenges environmental design for pervasive computing systems. *Proceedings of the Annual International Conference on Mobile Computing and Networking, MOBICOM*, 261–270. Association for Computing Machinery (ACM).

118. Hwang, K. (2007). Recent advances in trusted grids and peer-to-peer computing systems. *Proceedings—21st International Parallel and Distributed Processing Symposium, IPDPS 2007; Abstracts and CD-ROM*. IEEE Computer Society Technical Committee on Parallel Processing. DOI 10.1109/IPDPS.2007.370599

119. Nasution, M. K. M., Sitompul, D., & Harahap, M. (2018). Modeling reliability measurement of interface on information system: Towards the forensic of rules. *IOP Conference Series: Materials Science and Engineering, 308*(119). DOI 10.1088/1757-899X/308/1/012042

120. Chernov, V. M. (2019). Number systems in modular rings and their applications to "error-free" computations. *Computer Optics, 41*(5), 901–911. DOI 10.18287/2412-6179-2019-43-5-901-911

121. Awang Iskandar, D, N, F., Pehcevski, J., Thom, J. A., & Tahaghoghi, S. M. M. (2007). Social media retrieval using image features and structured text. *Lecture Notes in Computer Science (Including Subseries Lecture Notes in Artificial Intelligence and Lecture Notes in Bioinformatics, LNCS), 4518*, 358–372. DOI 10.1007/978-3-540-73888-6_35

122. Grabowics, P. A., Ramasco, J. J., Moro, E. Pujol, J. M., & Eguiluz, V. M. (2012). Social features of online networks: The strength of intermediary ties in online social media. *PLoS One, 7*(1). DOI 10.1371/journal.pone.0029358

123. Nasution, M. K. M. (2020). A method for constructing a dataset to reveal the industrial behaviour of big data. *IOP Conference Series: Materials Science and Engineering, 1003*(1). DOI 10.1088/1757-899X/1003/1/012156

124. van Zoonen, W., Verhoeven, J. W. M., & Vliegenthart, R. (2016). Social media's dark side: inducing boundary conflicts. *Journal of Managerial Psychology, 31*(8), 1297–1311. DOI 10.1108/JMP-10-2015-0388

125. Dini, A. A. (2017, August). The dark side of social media in eparticipation: A social-legal perspective. *AMCIS 2017—America's Conference on Information Systems: A Tradition of Innovation*. Americas Conference on Information Systems.

126. Nasution, M. K. M., & Noah, S. A. (2010). Superficial method for extracting social network for academics using web snippets. *Lecture Notes in Computer Science (Including Subseries Lecture Notes in Artificial Intelligence and Lecture Notes in Bioinformatics, LNAI), 6401*, 483–490. DOI 10.1007/978-3-642-16248-0_68

127. Nasution, M. K. M. (2016). Social network mining (SNM): A definition of relation between the resources and SNA. *International Journal on Advanced Science, Engineering and Information Technology, 6*(6), 975–981. DOI 10.18517/ijaseit.6.6.1390

128. Nasution, M. K. M., Hardi, M., & Sitepu, R. (2016). Using social networks to assess forensic of negative issues. *Proceedings of 2016 4th International Conference on Cyber and IT Service Management, CITSM 2016*. Institute of Electrical and Electronics Engineers Inc. DOI 10.1109/CITSM.2016.7577513

129. Nasution, M. K. M., Hardi, M., Sitepu, R., & Sinulingga, E. (2017). A method to extract the forensic about negative issues from Web. *IOP Conference Series: Materials Science and Engineering, 180*(1). DOI 10.1088/1757-899X/180/1/012241

130. Nasution, M. K. M., Elveny, M., Syah, R., & Noah, S. A. (2015). Behavior of the resources in the growth of social network. *Proceedings—5th International Conference on Electrical Engineering and Informatics: Bridging the Knowledge between Academic, Industry, and Community, ICEEI 2015*, 496–499. Institute of Electrical and Electronics Engineers Inc. DOI 10.1109/ICEEI.2015.7352551

131. Lubis, A. R., & Nasution, M. K. M. (2023). Twitter data analysis and text normalization in collectings standard word. *Journal of Applied Engineering and Technological Science, 4*(2), 855–863. DOI 10.37385/jaets.v4i2.1991

132. Lubis, A. R., Nasution, M. K. M., Sitompul, O. S., & Zamzami, E. M. (2020). A framework of utilizing big data of social media to find out the habits of users using keyword. *ACM International Conference Proceeding Series*, 140–144. Association for Computing Machinery. DOI 10.1145/3411174.3411195

133. Lubis, A. R., Nasution, M. K. M., Sitompul, O. S., & Zamzami, E. M. (2020). Obtaining value from the constraints in finding user habitual words. *2020 International Conference on Advancement in Data Science, E-Learning and Information Systems, ICADEIS 2020.* Institute of Electrical and Electronics Engineers Inc. DOI 10.1109/ICADEIS49811.2020.9277443

134. Lubis, A. R., Sitompul, O. S., Nasution, M. K. M., & Zamzami, E. M. (2022). Feature extraction of tweet data characteristics to determine community habits. *2022 5th International Conference on Computer and Informatics Engineering, IC2IE 2022,* 309–313. Institute of Electrical and Electronics Engineers Inc. DOI 10.1109/IC2IE56416.2022.9970180

135. Lubis, A. R., Nasution, M. K. M., Sitompul, O. S., & Zamzami, E. M. (2023). A new approach to achieve the user's habitual opportunities on social media. *IAES International Journal of Artificial Intelligence, 12*(1), 41–47. DOI 10.11591/ijai.v12.i1.pp41–47

136. Haggerty, J., Casson, M. C., Haggerty, S., & Taylor, M. J. (2012). A framework for the forensic analysis of user interaction with social media. *International Journal of Digital Crime and Forensics, 4*(4), 15–30. DOI 10.4018/jdcf.2012100102

137. Ashraf, N., Mahmood, D., Obaidat, M. A., Ahmed, G., & Akhunzada, A. (2022). Criminal behavior identification using social media forensics. *Electronics (Switzerland), 11*(19). DOI 10.3390/electronics11193162

138. Haggerty, J., Haggerty, S., Taylor, M. J., & Casson, M. C. (2013). A framework for the forensic analysis of user interaction with social media. *Emerging Digital Forensics Applications for Crime Detection, Prevention, and Security,* 195–210. IGI Global. DOI 10.4018/978-1-4666-4006-1.ch014

139. Lopes, M. E., Jacob, L., & Wainwright, M. J. (2011). A more powerful two-sample test in high dimensions using random projection. *Advances in Neural Information Processing Systems 24: 25th Annual Conference on Neural Information Processing Systems 2011, NIPS 2011.* Neural Information Processing Systems.

140. Rodriguez, C. E. (2022). Estimating the composition of the Chamber of deputies in the quick count for the 2021 federal election in Mexico. *Springer Proceedings in Mathematics and Statistics, 397,* 193–210. DOI 10.1007/978-3-031-12778-6_12

141. P´erez-P´erez, C. S., & Nieto-Barajas, L. E. (2022). Sampling design and poststratificaion to correct lack of information in Bayesian quick counts. *Springer Proceedings in Mathematics and Statistics, 397,* 163–176. DOI 10.1007/978-3-031-12778-6_10

142. Young, J., Yang, W., Young, S.-W., & Yu, Q. (2017). Presently untitled: Data mapping of 2016 U. S. presidential election twitter activity, phase III. *MM 2017—Proceedings of the 2017 ACM Multimedia Conference,* 580–581. Association for Computing Machinery, Inc. DOI 10.1145/3123266.3129332

143. Weyermann, C., Willis, S., Margot, P., & Roux, C. (2023). Towards more relevance in forensic science research and development. *Forensic Science International, 348.* DOI 10.1016/j.forsciint.2023.111592

144. Galante, N. Cotroneo, R., Furci, D., Lodetti, G., & Casali, M. B. (2023). Applications of artificial intelligence in forensic sciences: Current potential benefits, limitations and perspectives. *International Journal of Legal Medicine, 137*(2), 445–458. DOI 10.1007/s00414-022-02928-5

145. Bramley, R., Brown, A., Ellison, S., Hardcastle, W., & Martin, A. (2000). Qualitative analysis: A guide to best practice—forensic science extension. *Science and Justice—Journal of the Forensic Science Society, 40*(3), 163–170. DOI 10.1016/S1355-0306(00)71971-X

146. Mothi, D., Janicke, H., & Wagner, I. (2020). A novel principle to validate digital forensic models. *Forensic Science International: Digital Investigation, 33.* DOI 10.1016/j.fsidi.2020.200904

147. Ahmad, A., Diwan, S. P., Nanda, S. S., Sethi, L., & Patra, P. K. (2022). Digital forensic techniques and principles in a cloud environment. *Aiding Forensic Investigation Through Deep Learning and Machine Learning Frameworks,* 73–91. IGI Global. DOI 10.4018/978-1-6684-4558-7.ch002

148. Aboukadri, S., Ouaddah, A., & Mezrioui, A. (2023). Major role of artificial intelligence, machine learning, and deep learning in identity and access management field: Challenges and state of the art. *Lecture Notes on Data Engineering and Communications Technologies, 152*, 50–64. DOI 10.1007/978-3-031-20601-6_5

149. Nasution, M. K. M., Syah, R., Ramdan, D., Afshari, H., Amirabadi, H., Selim, M. M., Khan, A., Lutfor Rahman, M., & Sani Sarjadi, M. (2022). Modeling and computational simulation for supersonic flutter prediction of polymer/GNP/fiber laminated composite joined conical-conical shells. *Arabian Journal of Chemistry, 15*(1). DOI 10.1016/j.arabjc.2021.103460

150. Nasution, M. K. M., Elveny, M., Syah, R., Behroyan, I., & Babanezhad, M. (2022). Numerical investigation of water forced convection inside a copper metal foam tube: Genetic algorithm (GA) based fuzzy inference system (GAFIS) contribution with CFD modeling. *International Journal of Heat and Mass Transfer, 182*. DOI 10.1016/j.ijheatmasstransfer.2021.122016

151. Abraham, S., Alakananda, K., & Amdalli, N. A. (2023). A comprehensive review on digital forensics intelligence. *Advancements in Cybercrime Investigation and Digital Forensics*, 25–48.

152. Morelato, M., Cadola, L., B´erub´e, M., Ribaux, O., & Baechler, S. (2023). Forensic intelligence teaching and learning in higher education: An international approach. *Forensic Science International, 344*. DOI 10.1016/j.forsciint.2023.111575

5 Preventing Online Financial Frauds

Carrying Out Digital Forensic Investigation of Artificial Intelligence

*Raunak Sood, Anurag Sood, Shilpi Sood,
and Shubham Gajanan Kawalkar*

5.1 INTRODUCTION

Majorly, the contents will be dealing with taking a brief look at the personal data residing within mobile phones, so basically, there is a need to balance privacy and investigation when using digital forensics. When making ethical use of artificial intelligence and digital forensics, it is crucial to see that personal non-relevant data, when investigated, is not messed up by the concerned persons in charge of conducting a forensic investigation. Privacy-based ethical considerations majorly focus on privacy penetration testing, protecting the data against any third party gaining access to it, recovering data which is deleted, and incidental consent-based issues. One of the proposed solutions includes accessing control over the data by using techniques of secure erasure. An investigating agency should always prefer to make the digital forensics more privacy-friendly. Consequently, we will also delve into the issues and challenges connected with AI and digital forensic investigation, especially when dealing with the attribution of liability when investigating AI-enabled online financial fraud. Toward the end of this chapter, the authors will be proposing solutions for using digital forensics combined with artificial intelligence for doing away with online financial frauds, by using multiple training datasets and looking at the emerging technologies. Furthermore, AI principles and a working model of an AI bot will be discussed hereafter, dealing with how AI models related to fraud detection within financial channels.

5.1.1 USING THE POWER OF DIGITAL FORENSICS FOR COMBATING ONLINE FINANCIAL CRIMES

In this cyberworld, the persons who are engaged in perpetuating fraud have adopted different *modus operandi* to perpetuate cyber frauds. The world is being led by a

DOI: 10.1201/9781003501152-5

technological revolution, and with the changing technology, cybercriminals have also changed their techniques for committing frauds. It is now extremely essential to understand digital forensics to conduct investigation of online financial frauds in the vast cyberspace. *Digital forensics* is a branch of forensic science which involves the identification, collection, analysis, and preservation of digital evidence from various devices. Digital forensics is used for uncovering and unearthing, through careful investigation, data breaches and cybersecurity incidents, by using a variety of tools, techniques, and methods to unravel, give an interpretation, and finally, present the analysis report, in a manner which is admissible in the court of law [1]. According to Ombu, it is important that digital forensics must follow procedures during an investigation where the LEAs are able to unearth data which has been carefully concealed by the accused in hidden folders, retrieve deleted data, recover sabotaged and damaged files, and decrypt data using various forensic tools and techniques [2]. To simply explain online financial fraud, it is necessary to understand that *financial fraud* is causing wrongful loss to someone or wrongful gain to the benefit of another person. Analysis of digital evidence helps in proving the circumstantial link between a crime and the act or omission committed by the accused [3]. Majorly, evidence given by a forensic export is advisory in nature, and admitting it is completely upon the discretion of the court [4]. The canons of digital forensic investigation are briefly described as follows:

1. *Identifying the digital evidence.* In this step, it is important to understand what can be considered digital evidence at a crime scene—especially pen drives, hard disks, PDAs (personal digital assistants), tablets, and other electronic devices which might be able to store data (this storage includes internal as well as external storage) [5].
2. *Acquisition.* This stage comes after the stage of identification of relevant digital evidence. Now, the next task is to acquire data from the device by creating a forensic image of the identified digital evidence. To collect data out of the digital evidence, we must classify it further as volatile memory or non-volatile memory, because the acquisition method will differ for the varied type of memory. The main task of the forensic investigator is to acquire the data by employing a forensically sound method to make the forensic image of the original drive.
3. *Preservation.* The evidence which has been acquired or seized should then be transported to the FSL (forensic science laboratory) with a strictly documented chain of custody (COC) for preservation of evidence from being tampered either knowingly or unknowingly [6].
4. *Examination and analysis.* In this stage, the original evidence or seized device data (including the metadata) is analyzed by employing various forensic tools. Once the device has undergone a thorough examination, an analysis is done on the information collected via the various tools, by identifying the methodology and employing skills and tools for extracting vital information which can be used in the court of law. This is quite an important step because it is totally dependent on the skill and experience of the forensic examiner [7].

5. *Presentation*. This is the last step, where the forensic examiner of the electronic device presents all his findings in a well-documented report. The forensic examiner articulates his findings with the reasonings following through the legal issues and standard operating procedures and the extensively peer-reviewed methodology [8].

The LEAs are gearing up to battle the menace of online financial fraud, but there is a need to ensure that there is an existing mechanism to conduct a digital forensic examination of online financial frauds.

5.1.2 ANALYZING THE DYNAMIC NATURE OF FINANCIAL FRAUD AND USING CYBERSECURITY AND DIGITAL FORENSIC TOOLS AND TECHNIQUES AS ONE OF THE PROPOSED SOLUTIONS

Before delving into digital forensic techniques, it is crucial to understand the *modus operandi* of financial fraudsters, and the most employed modes of conducting financial fraud are [9]:

1. Phishing
2. Card skimming
3. Smishing
4. Vishing
5. SIM swap fraud
6. Identity theft [10]
7. And other incidental categories (refer to Table 5.1)

The common link/genus/connection of *modus operandi* among all these financial frauds is that the victim either knowingly or unknowingly or voluntarily gives out his sensitive personal information, which is extracted through deceptive practices employed by the fraudsters [11]. A study was conducted where the researchers carried out a bibliometric analysis of the user requests, where the main terms were "cybercrime," "online financial fraud," "fraud," and "finance fraud" [12]. These terms were taken out of scientific journals and writings which represented the collective requests of internet users. The result was that 35% of online frauds are completely "financial" in nature; therefore, dynamicity and natural pseudo-warfare are going on in the area of online financial fraud. It was also noticed that online financial frauds seem to be perpetrated more in developed and developing countries [13].

5.1.2.1 Obstacles and Issues Faced during Investigation of Online Financial Frauds

There are many obstacles which are faced during the traditional investigation of online financial frauds, one of the major concerns being "**anonymity**," which is allowing persons to participate in unlawful activities where they can protect their identities, acts, or omissions by using proxy servers [14]. These proxy servers are helping unscrupulous elements by masking their IP addresses and replacing it with a different IP address, thus making it difficult for the investigation to proceed further to pin liability on the

TABLE 5.1

Working Principles of AI for the Detection, Prevention, and Deterrence of Fraud

Type of Fraud	Usage of AI	Fundamental Idea (Detection/Prevention/ Deterrence)
Phishing	Using advanced ML algorithms for pattern analysis in communications which are taking place within the frauds reported and the investigation reports for the identification of phishing attempts. For example, by analyzing emails or suspicious headers, AI can send alerts to the user if there is a phishing attempt.	Detection
Identity theft	AI technology may be used in detecting the face of authorized person and then authenticating the identity of the person against the structured data retrieved from the identification cards of the person concerned. AI can be used to real-time verify the presence of the user through processing structured learning (verified documents, images) and unstructured learning (image and video).	Detection, prevention, and deterrence
Laundering money through the use of mule accounts (money muling)	AI can be used to monitor transactions by conducting pattern analysis of the potential laundering or fraudulent activities. AI can be used to flag the fraud activity using fraud flags (screening the sanctions lists and persons on watch lists).	Detection and prevention
Account takeover	To curb the menace of account takeover by fraudsters, AI-powered solutions may help in document verification, live check of identity, or authentication of the transaction using biometrics.	Prevention
Deepfake	AI can help in adding a layer of protection where the verification of identity takes place along with checking the liveness of the concerned person, by looking out for changing variations of the dynamic nature of human interaction.	Prevention

criminal. For example, Tor, I2P (Invisible Internet Project), and Freenet are the most common networks in the *modus operandi* of criminals who are misusing anonymity by encrypting incoming traffic and masking the IP address, thus engaging in the active concealment of their locations, website usage (by using the "hidden services" feature to maybe access the dark net), and internet-based activities [15].

Attribution of liability is another issue which needs to be dealt with in case of online financial frauds, because attribution becomes extremely difficult in cases of botnets, malware-infected computers, and the creation of background entry via the use of remote access tools (e.g., AnyDesk). In such cases, even if the device is physically in the control of a user, the user might not have the knowledge of what is there in his device, thus making it difficult to put together a chain of causation to attribute penal liability to an individual, because there might be an absence of *mens rea* altogether [16].

Traceback, or backtracking the cause to understand the nature of financial fraud, by investigating the devices by looking at event logs and incidental application logs, is quite an uphill task and time-consuming at the same time. This is especially an incredibly difficult task in case of analyzing the computers used to commit DDoS and DoS attacks against a single website [17].

Lastly, one of the major issues is the lack of uniformity of cybercrime-related laws across the world. The major lack is in evidentiary value standards (related to the admissibility and relevance of the digital evidence to the fact in issue), and the lack of mutual assistance treaties on cybercrime issues. A severe shortage has been observed in procurement, purchase, and capacity-building in the area of digital forensics. Legal issues are a major issue surrounding cybercrime which need to be addressed urgently [18].

5.1.3 PROPOSED SOLUTION FOR THE DIGITAL FORENSIC ANALYSIS OF ONLINE FINANCIAL FRAUD

By using tools like autopsy, it is easy to extract data from the device. The following steps can be followed to extract data from a forensic image:

1. Create a forensic image of the hard drive of the computer of the victim. The image can be created by using FTK Imager.
2. Once a forensic image is generated, then it is preferable to open that image in an autopsy software.
3. Add in a new case, and fill in all the details.
4. Once all the details are filled, then conduct a keyword search of the relevant terms.
5. Even the deleted files can be accessed by clicking on the View tab or when entering the relevant keywords deleted that may have access thereof.

5.1.3.1 Using AI Bots

It is suggested that the integration of AI bots with autopsy may be a potential solution, because the major function of the bot will be to retrieve emails by using the **confirmatory factor analysis** (an analysis conducted by observing the fit between the data which has been deduced via observation and, beforehand, conceptualized models—basically the training module of the AI) [19]. Then the data extracted may be classified as "further investigable" or "non-investigable" (through supervised deep learning, an AI bot can be prepared to enable the working of an AI-integrated autopsy framework) [20]. Then, when the AI retrieves the deleted data under the "further investigable" category, then tracking the IP address of suspicious emails and messages, or sending notices to the intermediaries, can take place, depending

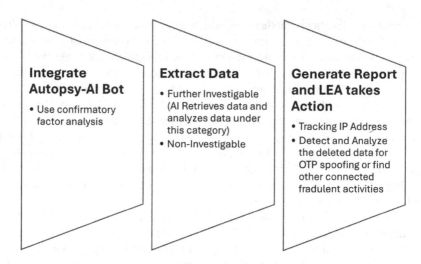

Integrate Autopsy-AI Bot

• Use confirmatory factor analysis

Extract Data

• Further Investigable (AI Retrieves data and analyzes data under this category)
• Non-Investigable

Generate Report and LEA takes Action

• Tracking IP Address
• Detect and Analyze the deleted data for OTP spoofing or find other connected fraudulent activities

FIGURE 5.1   Working principles of an AI bot.

upon the legal nuances which are to be taken care of by the LEA. The data extracted which might have been deleted might be able to help detect OTP spoofing and other mechanisms which are used to commit financial fraud. Then, the AI bot might help match the victim's stolen credentials and information which has been extracted from the device of the perpetrator. [21] Figure 5.1 shows the working principles of an AI bot.

5.2 USING ARTIFICIAL INTELLIGENCE TO DETECT, DETER, AND PREVENT ONLINE FINANCIAL FRAUD AND ANALYZING THE EMERGING TECHNOLOGIES

When using AI to detect online financial fraud, majorly, AI works on data analysis (using machine learning trained on datasets and analyzing datasets to analyze the common patterns and foresee the consequences), predictive modeling (using datasets to look for patterns), anomaly detection (raising red flags when real-time frauds might happen), and finally, biometric authentication (facial feature recognition and detection should be enabled to deter identity theft–based frauds).

5.2.1 AI MODEL FOR FRAUD DETECTION

Using data analytics and AI for detecting fraud helps in supporting internal audits. By using regression algorithms, decision trees, and other data from banks, it can be shown that processing large amounts of bank data using logistic regression algorithms proves to be far better than using other algorithms.

The major challenge for making an AI model for fraud detection is that fraud detection is extremely difficult, as the pattern of fraud is dynamic and changing. The perpetrators do not follow the same pattern always; hence, machine learning cannot rely on any specific training modules.

So hence, when making an AI model, the following factors need to be looked into:

a. The inputs/predicator and output (fraud labels)
b. The ML method for prediction
c. Criteria for the evaluation of or rating the fraud prediction model [22]

Looking at the "data" part, it is essential to take into consideration whether the data generated by the AI model is relevant or not. Firstly, many systems follow simple models to ML (machine learning), where it is necessary to discern the quantity of data (here, it is necessary to see the data as structured or unstructured). Secondly, the quality of data is dependent on the KYC conducted by the financial institutions. Now, there is trouble in integrating it with the ML model because of the difficulty in work-flow, data management, and change in controls, but using post analysis reports of transactions where money has been laundered can serve as good-quality dataset. ML can work on the post-analysis reports, and these can get integrated into the payment process systems through ML. Thirdly, mapping the data might help in facilitating the process, by the use of ERP systems to conduct fraud detection and prevention more efficiently. [23]

Supervised learning through the use of financial fraud statements can help in the generation of fraud labels and predictors, thus taking care of the dependent and independent variables. One of the best tools can be the use of available accounting data and standards to combine textual and network data together. To select the list of fraud predictors, it is essential to investigate the costs and benefits associated with it, especially the unstructured data consisting of audio; image is quite difficult to be engaged in. [24]

Algorithms. The **XGBoost algorithm** of ML is one of the popular algorithms particularly referring to gradient boosting, where majorly huge amounts of decisions are collected for the creation of an accurate fraud detection model wherein every decision tree will be used to correct the mistakes of the previous decision trees and works on the unbalanced data where approximately 1 in 1,000 transactions is turning up to be a fraudulent transaction. All decision trees will, at the end, contribute to the final result in the form of a score that will be created, indicating the probability of the transaction being fraudulent or not. [25]

5.2.2 CASE ANALYSIS OF EMERGING TECHNOLOGIES

1. *HyperVerge.* This is a next-generation AI-powered fraud detection system which helps with verification (document, liveness, biometric, and deepfake detection), authentication, and validation. It is a technology which detects blacklisted fraudsters and does not allow them to on-board the financial networks. Detection of any unauthorized identity is done on an immediate basis. Uses facial recognition. [26]

2. *Veritone iDEMS.* This is an extremely useful technology for LEAs. This technology works by using the aiWARE platform, where the user inputs

a huge amount of digital evidence and the platform manages the evidence via intelligent search and discovery by simultaneously ensuring that data is being shared and distributed across all the workstations. Massive amounts of digital evidence can be analyzed via Veritone technology by addressing the challenges of extremely high volume of digital evidence. It uses the redaction technology by tagging the sensitive personal information of users with the meta tags and ordinary tags, thus resulting in easy and efficient digital evidence management. [27]

5.3 ETHICAL CONSIDERATIONS AND PRIVACY-FRIENDLY DIGITAL FORENSIC INVESTIGATION OF AI

Daily, we are using devices like cell phones, PCs, and other smart devices in this age of technology. From the perspective of an investigator, data can be classified into relevant data or non-relevant data, whereas from the perspective of a user, the data can be classified as personal data and non-personal data; thus, this gives rise to four classifications, as follows:

1. Personal relevant data
2. Personal non-relevant data
3. Non-personal relevant data
4. Non-personal non-relevant data

During investigation of AI logs, the investigator has the right to access 1, 3, and 4 since the privacy laws do not protect non-private data or data which has been made public. [28] Thus, an investigative process can be quite intrusive of the privacy of a user, where personal non-relevant data is involved.

Privacy laws majorly follow the threefold test.

- *Legality.* Substantive and procedural laws both because, in forensics, we have to see whether we are violating the provisions of the DPDP Act or not.
- *Necessity.* Whether it is absolutely necessary.
- *Proportionality.*

Privacy laws are applicable to processing personal data in digitized form and have an extraterritorial applicability. The laws are not applicable to data made publicly available by individuals themselves.

5.3.1 PRINCIPLES ON WHICH THE PRIVACY LAW FOCUSES UPON

The privacy law mainly focuses on two major principles, namely, purpose limitation (processing of personal data for lawful purpose for which the individual or data principal has given consent, and the data principal should be aware of the purpose for which the data is being processed) and collection limitation (only the necessary amount of data should be collected for processing via lawful means). [29] In any forensic investigation, there are stages, like acquisition of data, preservation of the data collected, analysis of the collected data, and maintenance of continuity, to give an explanation in court and show how data has been handled, to disclose to the defense counsel in court so that the accused has a fair chance of protecting himself, and finally, to present all the collected evidence in court. [30]

5.3.2 Privacy Issues Regarding the Investigation of Online Financial Fraud

Whenever there is an investigation of a device, a forensic image copy of the device is made so that when the acquisition of data takes place, then there will not be a chance of any form of data alteration or modification. In due course of a forensic examination, it is not preferred to directly examine a device, because the data may get altered during the examination stage and all examinations taking place on the forensic image copy should not be affecting the original evidence in any manner. [31] Performing a forensic analysis of cell phones and smartphones is a challenge in itself because such devices are connected to the internet and are being constantly updated with messages, they receive phone calls, and the data sources include but are not limited to social media data, messaging services, notifications, and internet browsing activities. Receiving updates and the use of GPS might lead to alterations in data. Therefore, having a forensic image copy becomes essential for authentication, for continuity, and later on, for proving the integrity of the forensic examination. [32] During the examination stage, the privacy of the owner of the device becomes a point of contention, because the forensic team has access to all the contents of the device, so they index all the contents of the device, including the met data, and use digital fingerprints (a database of known "bad" files) to examine the files associated with the criminal activities. [33] Forensic examiners likely do not know who the owner of the device is; thus, one facet or angle to "privacy" is that devices are subjected to access controls which cannot be accessed via physical means, and sometimes investigators see totally non-relevant private information, which might become a cause of embarrassment for the owner of the device, because many people at large may come to know of the contents of the device when displayed during trial. The personal data of individuals is stored on PCs or on cloud services. The law provides exception for law enforcement, but in a cybercrime, there are many stakeholders, like the victim, the accused, and other third parties intentionally or unintentionally related to the crime; thus, to protect the privacy of such unrelated individuals, it is crucial that a victim consent form be required, to be filled out by the various other stakeholders who are not the accused. [34]

A broadly general summary of the privacy issues is presented in Table 5.2.

There are other privacy concerns, like the personal data of users being sold to third parties for processing, and even the privacy of local Wi-Fi networks is somewhat debatable because of Wi-Fi and smart devices and cloud computing, where the data generated in the cyberspace is stored. A device might relate to many other devices via local area networks or via virtual networks, so it becomes very difficult to pinpoint where the data generated is being stored, thus making it extremely difficult to establish a chain of custody.

Then there are infrastructural vulnerabilities, like smartphones being very small and compact in size, and if we put encryption into them, then the same might not be

TABLE 5.2

Digital Forensic Procedure for AI and Privacy Issue

Digital Forensic Procedure for AI	Privacy Issue
Just collecting or acquiring all the data without having due regard to what data is relevant or not.	There can be data in the device which might belong to third parties but is not relevant to the investigator; thus, there is a possibility that forensic analysis may intrude into the privacy of third parties who are unrelated to the investigation.
All deleted images in the disk are recreated and then they are analyzed.	Recovering such files might go against the scope of the investigation, thus intruding into the privacy of the individuals.
Data is being retrieved from the electronic devices which are undergoing investigation.	There needs to be unambiguous consent from the side of the user. There are lots of stakeholders not party to the investigation, but still, their data will be accessed without consent, which is a big privacy issue. For example, A is accused of cybercrime where B is the victim, and during investigation, it comes out that C might be a conspirator, based upon suspicion. Then the device of C can be seized, but C should fill out a consent form when submitting his device for forensic examination.
It might come to the knowledge of the investigator that the data is stored in servers of a foreign country.	Fragmentation of privacy law across the globe and lack of uniformity in global privacy law.
Digital forensics principles are guaranteed that they will be able to securely unearth the underlying criminal activity in the data present in the device.	In light of the forensic examination, the use of appropriate encryption when investigating data and integrating the same digital forensic methodologies and tools is of crucial significance.

feasible because it will affect the limited amount of computational capacity; thus, putting in extra security protocols into the device might prove to be quite difficult. [35]

5.3.3 MAKING DIGITAL FORENSICS MORE PRIVACY-FRIENDLY FOR AI WITH RESPECT TO ONLINE FINANCIAL FRAUD

1. *Database.* When the investigator collects data during investigation or during the course of an investigation, then major privacy problems, like protection of access to database, ensuring secure data erasure, and aligning with specific frameworks for storage of data, arise. The privacy problem has two stakeholders, namely, the server and storage device which has been connected to the device. The solution to this problem is tailoring the privacy policies to comply with storage limitation, purpose limitation, and adopting technologies for privacy, respecting investigations. [36] Two proposed technological solutions are:

a. *Access control.* Mainly used to protect privacy in the DNA databases of individuals. Here, the database consists of data of an individual which has been taken out from the device by the investigator. The solution is that the user and the investigator both put their heads together and classify data into various levels, where the user chooses between personal and non-personal data and the investigator chooses between relevant and non-relevant data, based on the scope of the issues under consideration for data analysis. So the proposed solution is that data should be encrypted wherever the user says that specific amount of data is private or personal, and the investigator also has a choice to discard the non-relevant data. But only one specific team of the investigator skims through the private data after, using very high amount of encryption. The main drawback is that encryption changes the nature of digital evidence which can be collected, but privacy law is very much flexible, as it allows "reasonable security measures"; thus, this could be a potential solution. [37]

b. *Secure erasure.* The main objective of secure erasure is to make sure that data, once erased, is not recoverable by any other means; thus, anti-forensics is one of the proposed solutions in this area to protect privacy. One method is to encrypt the logs with different keys and then, in the end, to just delete the keys altogether when there is no need to necessarily access the log records at all. [38]

c. *Framework and policies.* Adhere to GAPP (generally accepted privacy practices) and FIPS (fair information privacy principles) to make policies which are both business-friendly and that envisage that the investigations within the business organization are private and thus less-restrictive compared to public ones. The framework can be that there are three modules: (i) advanced system, (ii) evidence extracting module, and (iii) ranking. Herein, the evidence extracting module will collect the data out of the expert system, which in turn decides what type of digital evidence is relevant or not. But the main drawback is that the system mainly depends upon the quality and the quantity of the previous investigations conducted, and this drawback in itself is a privacy issue because there is a need to access case data without the consent of the user. [39]

d. *Start at manufacturing level.* Privacy and security policies have multiple stakeholders, like manufacturers, policymakers, and end users. Manufacturers can help in upholding privacy and security by increasing the security layers; at the same time, they can maybe take care of the software and hardware flaws. At the same time, they can also have numerous keys to protect the device, but when keys are demanded for investigation, then they should be handed over to the investigating agency. And at the conclusion of the investigation, they should delete the keys given by the manufacturer. [40]

Finally, it is proposed that the objective for investigators should be to catch criminals or attackers while complying with data protection laws; hence, it is argued that the analysis of data from physical security systems is one of the important focal points of an investigation. For the sake of a comfortable investigation, data taken from electronic devices should be anonymized, and investigators should use anomaly detection tools to de-anonymize the data without revealing the identity of honest users. [41]

One of the important points in computer forensics is to collect relevant data and ignore irrelevant data for the sake of privacy preservation. The collection of relevant data should take place by following various cryptographic techniques and complying with the privacy law and auditing the investigative process for privacy compliance. [42] At the same time, data should be categorized in different levels for undergoing different degrees of encryption. [43] The collection of data during investigation requires cooperation between the data owner and the investigating officer to determine privacy levels, because the investigator collects all the data on the device, so personal non-relevant data should fall into a category where the investigating agency can manage data in the following manner:

1. *Classification of forensic data and its data collection.* The data access possibility is that for personal non-relevant data and non-personal non-relevant data, there should ideally be no collection of such data by the investigating agency. There should be direct collection of personal relevant data, and as far as relevant personal data, the user here exerts a choice. [44]
2. *Privacy levels.* Those data which are non-personal but relevant should be accessed directly, and there is no need for any cryptographic methods to apply to it ("directly accessible data"). Data which is relevant and personal should be imaged and analyzed ("preserved-privacy accessible data"), and finally, any data non-relevant toward the investigation should not be accessed at all ("non-accessible data"). [45]

Now, for selecting personal relevant data, there is a need to conduct a pre-analysis process when making a forensic copy of the data which has been obtained. Normally, the process followed is to collect forensic data and then take it to the computer forensics laboratory (CFL), depending on the factors mentioned in the privacy law, which says the data owner may categorize his data as personal or non-personal and try to fit his data in one of the privacy levels present in the encrypted data process. [46]

5.3.4 RECOMMENDATIONS

There are many organizations which have defined privacy standards across the globe, but the most common standard has been developed by the ISO (International Organization for Standardization) and IEC (International Electrochemical Commission), covering the stakeholders of privacy and the principles and terminology which may be used. Experts are now developing ISO/PC 317, and a new standard, that is, ISO/NP 23485, is very much in the pipeline, ensuring compliance with multiple laws across the world. [47]

When digital forensics is being conducted on the device in issue, then to secure the privacy of the device, there should be a tight security around the room where the device is being kept, and "reasonable security measures" under the privacy law should include a measure where passing around of personal non-relevant data amongst the police network should be completely prohibited, to safeguard the privacy of the owner of the device. [48] There is a need to set out clear policies that devices should be retained only until there is a need to retain it for investigative purposes, and regular inspections should be carried out. Conduct bi-annual inspections of the forensic kiosks. Another proposed mechanism is that a senior police personnel of appropriate rank who is not involved in the investigation is to give authorization after evaluating explanations from the investigating officer as to why and which content of the data is necessary for it to undergo forensic examination. For urgent and emergency cases, forensic analysis authorization can be taken up to rule the possibility of unearthing personal non-relevant data from the forensic analysis. [49]

5.4 CONCLUSION

The way forward is to use protection, threat detection, and the development of a response strategy. The strategy should be divided into three parts:

1. *Before the compromise.* This step should incorporate digital risk protection with a scenario and system providing for warning or alerts on the data leakage when AI is being used for fraud detection and prevention. This step should include the preemption of the financial fraud via online means and the identification of past frauds, and if possible, dark web monitoring should also be done during this stage.
2. *During the compromise.* When the compromise takes place, or when fraudulent activity is detected, efforts should be made to block the mule accounts and identify the mitigation measures which should be taken further.

Then, to overcome the ethical issue of AI and the privacy dilemma, it is to be noted that consumers have shown a general lack of trust with regard to AI, and new technology always has had technological issues and distrust on the consumer side; hence, it is necessary to boost consumer confidence in resorting to emerging AI-driven technologies.

REFERENCES

1. Nance, K., & Ryan, D. J. (2011, January). Legal aspects of digital forensics: A research agenda. In *2011 44th Hawaii international conference on system sciences* (pp. 1–6). IEEE.
2. Ombu, A. (2023). Role of digital forensics in combating financial crimes in the computer era. *Journal of Forensic Accounting Profession*, 3(1), 57–75.
3. Parakh, S. C. (2011, January 1). Expert witness. *Indian Journal of Anaesthesia*. https://doi.org/10.4103/0019-5049.84839.
4. Ibid.
5. Taylor, M. (n.d.). *Digital evidence in cloud computing systems*. Elsevier. Retrieved February 29, 2024, from https://doi.org/10.1016/j.clsr.2010.03.002.
6. IEEE Xplore. (2016, April 1). *The role of digital forensics in combating cybercrimes*. IEEE Conference Publication. http://dx.doi.org/10.1109/ISDFS.2016.7473532.
7. Harbawi, M., & Varol, A. (2016). The role of digital forensics in combating cybercrimes. In *4th international symposium on digital forensic and security (ISDFS)* (Vol. 4, pp. 138–142). https://doi.org/10.1109/ISDFS.2016.7473532.

8. Lee, J. (2008, March). Proposal for efficient searching and presentation in digital forensics. In *2008 third international conference on availability, reliability and security* (pp. 1377–1381). IEEE.

9. Acharyulum, G. V. R. K. (2011). Information management in a health care system: Knowledge management perspective. *International Journal of Innovation, Management and Technology*, 2(6), 534–537.

10. Biros, David P., Mark Weiser, and John Witfield. (2007). Managing digital forensic knowledge an applied approach. In *Proceedings of the 5th Australian digital forensics conference*. Edith Cowan University.

11. A Booklet on Modus Operandi of Financial Fraudsters. (2023). *Reserve Bank of India*. Retrieved March 3, 2024, from https://rbidocs.rbi.org.in/rdocs/content/pdfs/BEAWARE07032022.pdf.

12. Yarovenko, H., & Rogkova, M. (2022). Dynamic and bibliometric analysis of terms identifying the combating financial and cyber fraud system. *Financial Markets, Institutions and Risks*, 6(3), 93–104. https://doi.org/10.21272/fmir.6(3).93-104.2022.

13. Ibid.

14. Alfred Ferdinand. (2023). *The challenges facing specialist police cyber-crime units: An empirical analysis*. www.unodc.org/e4j/data/_university_uni_/the_challenges_facing_specialist_police_cyber-crime_units_an_empirical_analysis.html?lng=en&match=The%20challenges%20facing%20specialist%20police%20cyber-crime%20units:%20an%20empirical%20analysis.

15. Maras, M.-H. (2014). Inside darknet: The takedown of silk road. *Criminal Justice Matters*, 98(1), 22–23.

16. Lin, H. (2017). *Attribution of malicious cyber incidents. Aegis paper series no. 1607 of malicious cyber incidents*. Hoover Institution.

17. Pihelgas, M. (2013). Back-tracing and anonymity in cyberspace. In K. Ziolkowski (ed.) *Peacetime regime for state activities in cyberspace international law, international relations, and diplomacy* (pp. 31–60). NATO Cooperative Cyber Defence Centre of Excellence.

18. United Nations Office on Drugs and Crime. (2013, February). *Draft comprehensive study on cybercrime*. https://www.unodc.org/documents/organized-crime/UNODC_CCPCJ_EG.4_2013/CYBERCRIME_STUDY_210213.pdf.

19. Barik, K., Abirami, A., Konar, K., & Das, S. (2022). Research perspective on digital forensic tools and investigation process. *Illumination of Artificial Intelligence in Cybersecurity and Forensics*, 71–95.

20. Prasanthi, B. V., Kanakam, P., & Hussain, S. M. (2017). Cyber forensic science to diagnose digital crimes-a study. *International Journal of Scientific Research in Network Security and Communication (IJSRNSC)*, 50(2), 107–113.

21. Koul, S., Raj, Y., & Koul, S. (2020). Analyzing cyber trends in online financial frauds using digital forensics techniques. *International Journal of Innovative Technology and Exploring Engineering*, 9(9), 446–451. https://doi.org/10.35940/ijitee.I7185.079920.

22. Baltrušaitis, T., Ahuja, C., & Morency, L.-P. (2019, February 1). Multimodal machine learning: A survey and taxonomy. *IEEE Transactions on Pattern Analysis and Machine Intelligence*, 41(2), 423–443. https://doi.org/10.1109/TPAMI.2018.2798607.

23. Bao, Y., Hilary, G., Ke, B., & Artificial Intelligence and Fraud Detection. (2020, November 24; forthcoming). Innovative technology at the interface of finance and operations. In V. Babich, J. Birge, & G. Hilary (eds.) *Springer series in supply chain management*. Springer Nature and SSRN. https://ssrn.com/abstract=3738618; http://dx.doi.org/10.2139/ssrn.3738618.

24. Amiram, D., Bozanic, Z., & Rouen, E. (2015). Financial statement errors: Evidence from the distributional properties of financial statement numbers. *Review of Accounting Studies*, 20, 1540–1593. https://doi.org/10.1007/s11142-015-9333-z.

25. Dalager, B., & Jensen, T. (2020, April 21). *Artificial intelligence prevents fraud*. KPMG. https://kpmg.com/dk/en/home/insights/2020/04/artificial-intelligence-prevents-fraud-.html.
26. Hyperverge: The New AI-Powered Fraud Detection Technology. (2024, February 20). Retrieved March 29, 2024, from https://hyperverge.co/in/blog/how-to-leverage-ai-to-prevent-fraud-a-deep-dive-for-financial-institutions/#:~:text=AI%2Dpowered%20 fraud%20detection%20models,of%20identity%20fraud%20and%20deepfakes.
27. Ibid.
28. Careless, J. (2024, March 12). *How law enforcement can surf the rising tide of digital evidence without drowning in detail*. Police1. www.police1.com/police-products/ investigation/evidence-management/how-law-enforcement-can-surf-the-rising-tide-of-digital-evidence-without-drowning-in-details.
29. The Digital Personal Data Protection Act, No. 22 of 2023, S. 3(c)(ii). The Act does not apply to data which has been made "publicly available" and for "personal or domestic purpose" by the user and this term "publicly available" is broad enough to incorporate the non-private data of the user on public platform whereas the term "personal or domestic purpose" is broad enough to include the data stored on device which is non-private in nature as far as the concerns of the user are concerned.
30. Rustad, M. L., & Koenig, T. H. (2019). Towards a global data privacy standard. *Florida Law Review*, 71, 365.
31. Badiye, A., Kapoor, N., & Menizes, R. (2023). Chain of custody. In *Starpearls* [Internet]. Retrieved September 6, 2023, from www.ncbi.nlm.nih.gov/books/NBK551677/.
32. DIS0017—Evidence on Disclosure of Evidence in Criminal Cases. data.parliament.uk/ WrittenEvidence/CommitteeEvidence.svc/EvidenceDocument/Justice/Disclosure%20 of%20evidence%20in%20criminal%20cases/written/80580.html.
33. Ibid.
34. Sommer, P. (2019, May 24). *View my complete profile*. Peter Sommer: Digital Evidence and Policy. pmsommer.blogspot.com.
35. Ibid.
36. Ray, I., & Shenoi, S. (eds.). (2008). *Advances in digital forensics IV* (Vol. 285). Springer Science & Business Media.
37. Baye, B. C. (2022). *Exfiltrated data and the privacy laws digital forensics examiners need to understand for analysis* (Doctoral dissertation, Utica University).
38. Lytle, A., Stephens, N., Conner, J., Bashiri, S., & Jones, S. (2016). *Digital forensics and enforcement of the law*. Retrieved August 20, 2024, from https://internetinitiative.ieee. org/home/sitemap/20-newsletter/396-digital-forensics-and-enforcement-of-the-law.
39. Mazurczyk, W., Caviglione, L., & Wendzel, S. (2019). Recent advancements in digital forensics, part 2. *IEEE Security & Privacy*, 17(1), 7–8.
40. Årnes, A. (ed.). (2017). *Digital forensics*. John Wiley & Sons.
41. Montasari, R., Carroll, F., Mitchell, I., Hara, S., & Bolton-King, R. (eds.). (2022). *Privacy, security, and forensics on the internet of things (IoT)*. Springer.
42. Caloyannides, M. A. (2004). *Privacy protection and computer forensics*. Artech House.
43. Mitrakas, A., & Zaitch, D. (2008). Law, cybercrime, and digital forensics: Trailing digital suspects. *Information Security and Ethics: Concepts, Methodologies, Tools, and Applications*, 1681–1700.
44. Horsman, G. (2022). Defining principles for preserving privacy in digital forensic examinations. *Forensic Science International: Digital Investigation*, 40, 301350.
45. Ahmed, A., Javed, A. R., Jalil, Z., Srivastava, G., & Gadekallu, T. R. (2022). Privacy of web browsers: A challenge in digital forensics. In *Genetic and evolutionary computing: Proceedings of the fourteenth international conference on genetic and evolutionary computing, October 21–23, 2021, Jilin, China 14* (pp. 493–504). Springer Singapore.

46. Englbrecht, L., & Pernul, G. (2020). A privacy-aware digital forensics investigation in enterprises. In *Proceedings of the 15th international conference on availability, reliability and security* (pp. 1–10). https://doi.org/10.1145/3407023.3407064.

47. Dehghantanha, A., & Franke, K. (2014, July). Privacy-respecting digital investigation. In *2014 twelfth annual international conference on privacy, security and trust* (pp. 129–138). IEEE.

48. ISO. (2018). *Data privacy by design: A new standard ensures consumer privacy at every step*. Retrieved September 6, 2023, from www.iso.org/news/ref2291.html.

49. Nance, K., & Ryan, D. J. (2011, January). Legal aspects of digital forensics: A research agenda. In *2011 44th Hawaii international conference on system sciences* (pp. 1–6). IEEE.

6 Path to Intellectual Revolution in Digital Forensics

Author-Keerti

6.1 INTRODUCTION

6.1.1 THE SIGNIFICANCE OF DIGITAL FORENSICS

Virtual forensics is particularly concerned with the detection and prevention of cybercrimes. Each virtual forensic technique relates to virtual security as it focuses on virtual events. At the same time as virtual protection focuses on preventive measures, digital studies pay attention to remedial measures. Matthew N. O. focuses on virtual safety as he pays attention to virtual events. At the same time, he emphasizes preventive measures for virtual protection, while also giving attention to remedial measures in digital studies. It smiles from the paintings of locating evidence from digital environments along with computers, mobile phones, servers, and/or the internet. It offers forensic groups with the excellent era and equipment to resolve complicated digital-related troubles. The virtual forensics team examines, analyzes, and preserves digital proof discovered on numerous digital devices. In this way, law enforcement catches criminals and brings them to court. Forensics facilitates companies to recover, perceive, and preserve computers and associated facts because it helps them conduct investigations as witnesses in court dockets. It allows the prediction of the motive at the back of the crime and the identification of the actual criminal. The procedure of creating a suspect crime machine helps prevent the destruction of digital evidence received. Information series and replica: get better-deleted files and deleted partitions from media to extract and take a look at the evidence. It helps you speedily discover evidence and additionally allows you to estimate the chance of harm to the sufferer. *Behavior* is a PC forensics report that provides a duplicate of an entire report of the hunt system. See evidence of the following chain of custody.

6.1.2 AUTOMATED LOG ANALYSIS

Security teams often deal with large amounts of data generated by various systems, applications, and network devices, but reviewing these logs can be a time-consuming and error-prone experience. This is where automatic engine analysis comes into play. Artificial intelligence algorithms are good at processing large amounts of data,

DOI: 10.1201/9781003501152-6

analyzing its patterns and inconsistencies. Thanks to artificial intelligence–supported wheel analysis, investigators can look for suspicious activities, security issues, and areas that require further investigation. [1] Artificial intelligence increases the speed and accuracy of log analysis, allowing researchers to focus on areas of interest without wasting time and resources on manual analysis. [1]

For researchers, the author divides the analysis into two broad groups: There is data from network and security devices. Rephrase routers, switches, IDS, firewalls, proxies, WAF, etc. The text is from the edges. The method falls into two groups.

6.1.3 GENERALIZED COMPUTATION

This group refers to algorithms. It is designed to hit upon uncommon styles in string data. Two famous forms in this category are linear guide vector machines (SVM) and stochastic forests. For example, SVM shows the probability that certain words in a column are associated with a particular event. Certain words, such as "error" or "failure," will be associated with the result and will score higher than other words, such as "success" or "link." Useful words to describe the questions. The training of both SVM and stochastic forest models involves supervised machine learning and the use of large amounts of data to achieve accurate predictions. As we discussed previously, this makes it difficult and expensive to ship them to many parts of the world.

6.1.4 DEEP LEARNING

Deep learning is a type of machine learning commonly known as artificial intelligence (AI). Deep learning discovers patterns in data by training neural networks on large datasets, but it often combines with supervised training using the data. Researchers have used artificial intelligence with significant results to solve complex problems, such as image and speech recognition. The University of Utah's "DeepLog" article is one of the best studies on this subject. Their method uses deep learning to identify anomalies in logs. Interestingly, following the discussion, they also used machine learning to identify events similar to Zebrium, because this increased the accuracy of detecting anomalies.

The difficulty with this method is that it requires a lot of data to be accurate. This means that a new environment will take longer to provide an accurate prediction, and a small area will not produce enough data for the model to be accurate enough. Also, unlike the statistical algorithms discussed previously, deep learning requires extensive computational training. Many data scientists use expensive GPU techniques to train their models faster, but at a high cost. This can be a very expensive way to do engine analysis, because we need to train the model on each feature separately and continuously. Some vendors already provide third-party services (like MySQL). This approach works because it can use a lot of data and errors to train the model, and many users can use the training model. However, since few websites use only third-party services (and often proprietary software), this model detects events only from third-party services and not from software developed in the environment.

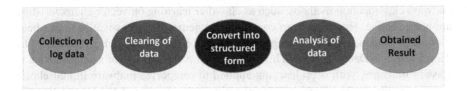

FIGURE 6.1 An automated approach to analyze logs using AI.

6.1.5 LOG ANOMALY DETECTION

Many motion study methods focus on anomaly detection. There are some problems with this: the daily volume increases, the engine runs noisily, and most of the time, there are no problems. This makes wheel abnormalities difficult to identify. It is very important to remember that rarely anything happens when solving the aforementioned problems. But when all events are not of the same type (they are part of events and are different), how do you know if the event is rare? At the very least, machine learning should be able to sort engine conditions by type to determine which ones are bad. The most common method is the longest common subsequence (LCS), but the difference between individual states of the same type makes the accuracy of LCS difficult when using logs. Log anomaly detection is often very popular (which increases classification error). Most logs have many flaws, and only some are useful for finding and/or solving problems. Therefore, experts need to examine themselves to find the noise and identify the suspect. Successful machine learning–driven analysis requires more than error. Machine learning for log analysis is a way to analyze logs using ML/AI (shown in Figure 6.1).

6.2 MALWARE DETECTION

Malware analysis is the process of analyzing malware to understand its functionality, capabilities, and impact. It is an important part of digital science because it helps researchers find stops, determine the extent of damage, and develop repair strategies. [1] The rapid evolution of malware requires detection technology. [2] AI-powered malware detection machines use learning to analyze and learn from user behavior to better identify malware and help researchers remove malware from infected networks. For instance, security companies. Artificial intelligence tools can be used to learn from known patterns and characteristics of malicious software. They can use big data to detect and classify new and previously unknown threats. There are similarities to previous threats and attacks.

Computer security faces a significant challenge with malware. Why? Every day, the AV CHEK Institute, an antivirus organization, reports the production of more than five million malware samples. Due to the inability of safety groups to handle all malware immediately, a malware class method is often necessary to prioritize specific incidents. The scope, magnitude, and sophistication of malware are rapidly expanding. Computer hackers and attackers create structures using algorithms that can automatically rearrange the code and encrypt it, evading detection. Traditional malware classification methods, such as classifier learning on vector characteristics, are not effective. However, deep convolutional neural networks (CNNs) have been successful in identifying and categorizing malware. This is further demonstrated by deep learning's Deep Desktop. This study recommends the use of a smart device to classify malware, with deep learning applied to categorize malware in households and into multiple classifications. Additionally, it emphasizes data visualization. To evaluate the effectiveness of our approach, we utilized a Microsoft malware dataset of 10,000 samples with nine different classifications. We then used overlapping neural networks to transform the document into a grayscale image and categorize the photo using convolutional neural communities that use fuzzy algorithms. The outcomes stood out in comparison to deep mastering models that were 99.97% accurate for nine malware categories.

On the World Wide Web (WWW), people consider malware as one of the biggest threats to system protection. Numerous types of malware, including viruses, worms, Trojans, botnets, and rootkits, penetrate and disrupt networks and systems, causing a combination of attacks. Malware regularly uses felony protection mechanisms to avoid detection. [3]

In addition, the outcomes of investigations into these assaults are regularly ineffective, and the lack of appropriate and immature equipment will make it difficult to achieve real evidence.

Forensic strategies. This study addresses the various demanding situations researchers face whilst detecting and analyzing malware (shown in Figure 6.2). In this text, the authors describe the want for a new technique for malware detection specializing in a study framework and providing answers based totally on complete fact analysis and marketplace research.

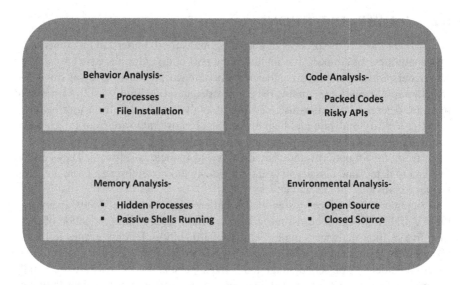

FIGURE 6.2   Four focus areas of malware analysis in digital forensics.

6.3 IMAGE AND VIDEO ANALYSIS

Analysis of images and videos is an important part of numerical analysis. For example, intelligent algorithms can scan large amounts of multimedia content and quickly identify faces, objects, or text in photos and videos, thus speeding up the search process, extracting key evidence, and supporting multiple investigative scenarios. [4] Consider a case where detectives must identify a suspect caught on a security camera in a crowded area. Reviewing videos is often complex and can take several hours. AI-powered facial recognition technology can quickly analyze large amounts of video data to identify persons of interest and reduce the amount of manual work required. Technology speeds up the identification process, allows investigators to focus on what matters most, and speeds up investigations. Video/image analysis involves the scientific study of analog or digital multimedia and video files, or printed or digital images, to improve the visibility of specific features contained in the video or image files or full or image files. [5]

The entire process of photo and video forensics is explained here. Forensic video analysis is broad and complex, just as computer forensics involves more than just copying and viewing data; many steps are often overlooked and rarely taken into account. If we consider all the work involved in the analysis, we will think that this is very difficult. As a video forensic analyst, you must consider it important to understand all the steps required to complete the actual analysis. This way, you can stay steady and reduce the possibilities of passing or lacking. Moreover, if you should go to court docket, you may have a report—a good way to form the premise of your presentation. It is more complicated than that. You need to describe the product, make the right decision, write down the process, compare it with other products, and then go to court. Since a digital file is just a collection of objects, here are the details of working with objects and what you should do with them. [5, 6]

Step 1: Recover data and get the product.
It can be very difficult and important for the integrity of the step. Unfortunately, analysts or people who understand the importance of being careful when collecting data do not take these steps. In most cases, primary evidence is collected not by the police investigation team but by police officers who first arrive on the scene, and low-level evidence or information is often sent to a disk or flash drive with limited resolution. First DVR record for form shooting the next day. This happens all too often and gets everyone off to a bad start. [6] Another way to get information is to leverage your colleagues in the field of digital forensics. If they are careful and take the necessary steps to document their process, you are off to a good start. Usually, analysts start burning videos to DVDs, or some images received by email, but we should not forget to write the previous steps. For example: analyzing disk images to obtain completed and deleted files, exporting videos to a DVR, copying files from a hard drive, back analog to back of videos (e.g., VHS hash codes, etc.). Investigating officers often use evidence initially collected by non-professionals, who may not follow best practices. However, the situation can be improved with better communication, education, and

diplomacy. There is a possibility that this is true. Let them know how you save and write. Overwriting the hard drive can be avoided. Explain to them how to export. Remember the make and model of the DVR if you want to see the poor-quality re-encoding. It is important that you make sure that if they have questions, they call you.

Step 2: Determine the information: What is this product?

Now that you have these items, you should take something from them to review. This is not a big problem if you have some normal video files and have the appropriate codecs. However, this can become a challenge in itself when there are hundreds of video file owners. You need to know what you are looking at to decide what to do. These steps generally describe the challenges you will encounter:

- You have a disk image and you need to recreate photos and videos, even if they have been deleted. (This happens, for example, with nude photos of children.) [6]
- You have a dump on your DVR drive but you do not know how to use it.
- You have a video trace export but the video is in a special format (very rare). Video players provided by manufacturers are often buggy, unusable, and incompatible with modern versions of Windows, and videos do not export correctly.

From this step, you may want to do a little study to realize exactly what you need to do. Do you want recognition on records restoration first? Need to discover a higher codec? Need to be a higher videographer, or discover a better way to send your video? On occasion, this is a clean step; however, occasionally, it is miles more tough and time-consuming. Keep in mind that that is a scientific observation, and occasionally, studies cannot be rushed.

Step 3: Find useful information: Where is the good information?

In this step, you should look at the video or picture, but find the right one! Here are two examples you will encounter:

- Find images that interest you in a large database.
- Find events that interest you during the video.

These steps can be done by other team members. Communication is supported by members working on the case. In many cases, the important thing you need to know is what happened and when. Of course, technologies like video content analysis and facial recognition can also help.

Step 4: Find the source: Where does the product come from?

You need to understand how the original data was created, depending on the situation. In a sense, we can call this ballistics. Knowing the type and location of the file can help you understand something about the case. Some tests that may be done are:

- Identify the type of source (digital camera, scanned image, computer-generated, etc.).
- Indicate the model of the camera used to capture the photographic image.

You need to fill out the base before you can preserve the evidence. This will help you if you need to go to court.

Step 5: Check the accuracy of the information: Is anyone interested in the product?

At this point, you may want to know how reliable the information you have collected is. Is it possible that someone changed it? This can be done at various levels:

• Indicate that the data has been edited (for example, change metadata).

• Indicate that the image has been edited (for example, change the format, resize, or crop).

Research. Although the content has been edited, tampering, such as removing or adding content, has become more common. [6] This can be done unintentionally (such as changing the format from raw text to non-resolution media) or deliberately "Photoshopped" to preserve originality. In this digital age, this is a problem that needs to be solved. You should always use raw data if possible, and this can be easily verified using forensic tools, such as Amped Software's Amped Authenticate.

Step 6: Evaluate data quality: Do you have enough seats?

At this step, you can see something in the picture, but you have to know if it is good enough for your purpose. Can you read a car's license plate? Do you have enough space for the recognition of a face in pixels? Does the image have the information you need? Is it possible for the information to be better recovered or viewed through an image enhancement? Is it re-establishment? If there is something wrong with the picture, is it possible to get it back?

Knowledge and experience are important to determine if you are good enough. It is not always easy. The minimum value is required to get good results. You can use the shortcut for things like faces and license plates. You have to zoom in and count the numbers. If you only have a small amount of time, you can draw all the characters on the plate. Very low efficiency is what it is. According to our experience, if the vertical resolution is less than 12 in, it is generally not a big deal; it is not possible to get information from the drive plate.

Step 7: Set up the equipment: take out the good stuff!

Once you identify the issue affecting your photo or video, you can enhance and edit your file with the right tools, like Amped FIVE from Amped Software. This step is very large and may include the following processes: contrast, equalization of the histogram, sharpening, image correction techniques (understand the mathematical model of information interference, and try to modify the model to restore the false image [remove blur]), Fourier filtering, and frame integration.

For hard-core video development for people, this is the most fun part of the process. It involves some trial and error, but at least it is fun. While no one can guarantee good results using the Hollywood magic of *CSI* shows, you may see some good results.

A very important thing to remember in this step is to document the development process so that you have a research document that can be presented to the court. Check all the procedures you have done.

Step 8: Analyze and compare data: What does this product represent?

Here you can see what you are doing. If the visual content is not developed in a way that you can understand and share, the improvement steps will be useless. In this step, you can do the following:

- Compare a face in two different photographs.
- Compare a face with an experience.
- Read the license plate permit of a car.
- Specify the location of this image.
- Measure the height of the object.
- On file, find the corresponding finger.

If you do not get the results you want, you can go back and repeat the steps (shown in Figure 6.3). [6] For this reason, you determine that the information you need cannot be obtained from the information provided. [7] Remember, this is a scientific process and can easily be distorted by excessive and inefficient work, but sometimes this is the card that pays off.

6.4 NATURAL LANGUAGE PROCESSING

NLP involves calculating and understanding human speech. This AI tool has advantages in digital forensics, especially when it comes to data crime. AI applies NLP to data and detects sensitive data, such as credit card information or personally identifiable information (PII), in unsecured domains. Social media evidence is a brand-new subject matter in digital forensics. Investigating social media posts can provide precious support to the research of many crimes if done efficaciously. Looking for facts on social media to provide crook evidence to the government is no smooth venture. [8] Virtual technology research is based on natural language processing (NLP) generation, and blockchain technology is planned in the system. The primary reasons for the usage of NLP in this method are record evaluation, representation of every level, stage vectorization, function selection, and type analysis. The application of blockchain generation on this machine can protect records from hacker assaults and network attacks. The device's abilities are tested using real international information.

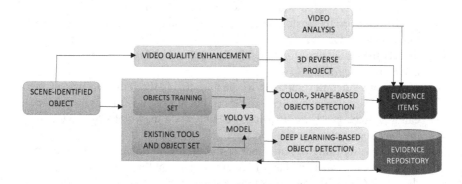

FIGURE 6.3   Video-based evidence analysis and extraction in digital forensic investigation.

Artificial intelligence technologies such as natural language processing (NLP) can identify important information from big data. For example, text documents such as emails, chat logs, and documents often contain valuable evidence in digital investigations. Use AI to discover relationships, uncover patterns, and analyze images better and more accurately when searching for text. [7, 8] Imagine a situation where investigators analyze large amounts of chat information to identify individuals involved in cybercrime. Natural language processing algorithms based on artificial intelligence can quickly process and analyze data to detect repeated words, suspicious patterns, and human relationships. This allows investigators to identify persons of interest and disclose confidential communications, speeding up the investigation process and allowing for timely intervention. For example, NLP technologies known as tokenization, preprocessing, provenance, and entity recognition (NER) can help remove relevant information from digital evidence, again without much trouble.

6.4.1 NLP Applications

- *Information extraction.* Extraction of statistics created from unevaluated facts.
- *Get entry to truth (ir).* Get percentages and essential facts or facts that are thrilling to mind.
- *Call entity popularity (ner).* Understand and retrieve entities with names, places, locations, and dates in the text.
- *Set content material records format.* Arrange statistics via categories or tags.
- *Grammar.* Test the text for footnotes and definitions.
- *Summary.* Create a summary of a protracted textual content. [9]
- *Translator.* Translate text from one language to some other.
- *Chatbots.* Chatbots are interactive tools that work well every day. People love them because they help them complete these tasks quickly, so they can focus on high-level, creative, and engaging tasks that require human abilities that machines cannot replicate.
- *Sentiment analysis.* Sentiment evaluation (or sentiment mining) is a natural language processing (NLP) technique used to decide whether or not a product is right, horrific, or impartial. Sentiment analysis is often executed by studying records to assist businesses in tracking merchandise and evaluations in client comments and understanding consumer desires. [10]

Figure 6.4 shows the applications of NLP. Over the years, digital evidence has increased not only in size but also in quantity: we now store more information on our computers and smart devices than ever before. With audio and video data becoming more common than ever, the use of various types of media has also increased. This creates a problem for digital analysts, who now have to examine these files and listen to their content. Smart models and applications that can be run locally provide new ways to decipher this complex evidence, reducing the time it takes to find key information in that evidence.

Applications of NLP (Natural Language Processing)

- **Email filtering**
- **Chatbots**
- **Sentiment Analysis**
- **Social media monitoring**
- **Predictive Text**
- **Smart Assistant**
- **Automatic Summarization**
- **Language Translation**
- **Online searches**

FIGURE 6.4   Applications of NLP.

6.5   NETWORK TRAFFIC ANALYSIS

Monitoring and analyzing network traffic patterns are crucial to diagnosing and responding to network attacks. Rather than manually inspecting and analyzing networks on a predetermined basis, forensic teams can train AI algorithms to obtain network data analysis, identify differences in normally operating traffic patterns, and sound alerts when the suspect needs further investigation. [1] AI can also help correlate cyber incidents to known attack patterns, providing incident response teams with better information.

Network traffic analysis can identify malicious behavior against specific IPs and can also be used as a diagnostic tool to determine how threats move laterally within an organization and to let you know what other devices may be infected. This enables faster response, preventing business interruptions.

6.5.1   IMPORTANCE OF NETWORK TRAFFIC ANALYSIS

Companies rely on NTA to improve network management and security. Here are some reasons organizations should adopt this in 2023.

1. **Automatically Identify Network Anomalies**
   Traditional solutions alert users when a device is not allowed to connect to the network or when the network behaves abnormally, but they do not consider replacing network valves. Network monitoring tools, on the other hand, do not require network security professionals to monitor DNS and DHCP information [11] and adjust management information, information services, and other information to get a complete view of traffic patterns and network performance. Alternatively, cybersecurity experts rely on networking tools for threat analysis to quickly find suspected vulnerabilities. Using

the information provided, they can find the root cause of the problem and provide a solution quickly.

2. **Network Always Available**

Every enterprise's community wishes to be to be had 24/7. Community visitors' evaluation presents corporations with information, approximately, community uptimes, at the same time as evaluating community connectivity and enhancing average network overall performance. Community element evaluation indicates harm to the subnet due to lack of connectivity, community interface failure, and terrible interference. In community time, network evaluation helps your protection crew identify threats and quickly discover the foundation motive of network failure, decreasing the probabilities of malicious customers.

3. **Strengthen Network Security**

Over the past few years, the number of online victims has been higher than ever before. Cybercriminals have perfected the art of attacking quickly without being detected. They use genuine credentials obtained through dubious methods, such as phishing, to gain access to trusted online sites.

Because they use legitimate access credentials, it is difficult for a network without access control to detect and block access. However, cybercriminals' efforts will be in vain if you use effective tools to monitor your network traffic. There is another reason that using a network analytics solution is important.

4. **Powerful Insight**

Due to the pandemic, many organizations have become relatively reliant on cloud computing, DevOps, IoT, etc. in their daily operations. Combined with remote operation, it is difficult for network administrators to maintain the network effectively. With NTA, the technology team can now solve the problem of lack of network visibility, because it is the organization's only point of truth. Simply put, NTA can create insights that no other source can provide.

Your technology team can easily use network connectivity analysis to understand your company's communications infrastructure. NTA provides real-time monitoring of TCP/IP packets, cloud transactions, API calls, vSwitch-based virtual network traffic, serverless events, and all network communications.

It also provides comprehensive information regarding network operations and IT team problem resolution. IT professionals can determine network locations and locations with multiple devices, create accurate top-down maps, improve network visibility, prevent blind spots in network content, and easily resolve the location.

5. **Network Performance**

High-availability networks must operate at maximum capacity. But to achieve this, your IT team must be able to track the performance of the resource by monitoring its usage. Network traffic analysis can help your team do this by supporting network capacity and planning.

NTA can help your team identify network connections that need to be upgraded, because it can find bandwidth issues. Your IT professionals can use NTA to identify network resources that may be interrupted and help reduce IT costs.

6.5.2 Network Traffic Analysis Features

Since there are many NTA solutions for data collection and security management for your business, you must use the best one. Thanks to network traffic analysis tools, you can constantly monitor your network and detect threats that may cross network boundaries or originate from businesses. These features will help you evaluate appropriate use.

1. **Built-in Threat Intelligence**
   Solutions with integrated analytics help your team detect malicious behavior on your network. Effective NTA tools have different names and methods designed to model behavior and data processes. This increases the accuracy of alerts, speeds up threat detection and response, and simplifies operations.
2. **Service of Key Metrics**
   The tool should be able to analyze data from large sources, including different times, bandwidth usage, and traffic of an application, to gain insight into network operations. If a device does not have this feature, then it is outdated.
3. **Understanding the Cloud**
   The device must be able to monitor cloud traffic. In the past period of KEVID, many online activities took place in the cloud. To stay relevant, your organization must be able to manage virtual private cloud infrastructure, application programming interfaces (APIs), and cloud monitoring engines for end-to-end visibility.
4. **Actionable Analytics**
   Although network traffic analysis tools are supposed to detect threats, it does not stop there. Your thoughts on security tools' answers should provide actionable recommendations for managing problems and provide solutions for analysis models for network security, bandwidth optimization, and prevention of future data breaches.

6.5.2.1 How to Develop Network Traffic Analysis

There are several steps you need to take when implementing NTA solutions that suit the desired needs of your company.

Step 1: Identify network resources.
Perceive and allocate community sources for your agency which are to be had for evaluation, along with computers, applications, servers, switches, firewalls, and routers. Those all offer exceptional metrics that may be analyzed. While there are manual and automated techniques of doing this, the automatic technique takes less time. It is also less complex because it makes use of automation and network discovery, together with SNMP, windows control, instrumentation (WMI), streaming protocol, and method tracking.

Step 2: Select the data source.
Pick out statistics from a dealer or without aggregation. Use software to extract data from analyzed information to use the agent approach. Computerized poor detection is brilliant for accumulating certain facts but can be the

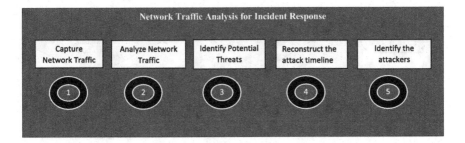

FIGURE 6.5 Investigating attacks with IPS and network forensics.

reason for storage and processing issues. However, with the use of the agentless method at the network layer, you use processes, techniques, and APIs that are already supported with the aid of the material files for your company. You can use SNMP on a network tool or WMI on a Windows Server.

Step 3: Data sampling.

Start by sampling lots of data; this should select many venues that provide diverse datasets for large businesses. This way, you can identify the problem at a small scale and then expand your entire network analysis to collect information from across your entire network. [11]

Step 4: Continuous checking.

Install continuous tracking and select the favored goal for statistics series. This permits you to come across protection threats, inspect failures, and benefit from long-term insights from historical statistics approximately for your solutions. It also permits your IT group to quickly and efficiently examine your community and locate breaches earlier than they manifest. Figure 6.5 suggests investigating assaults with IPS and community forensics.

6.6 FORENSIC TRIAGE

Forensic triage is the process of examining a computer to eliminate it or include it in a full analysis. Digital investigations contain a wealth of information and require investigators to quickly review and prioritize relevant evidence. [12] AI in forensic triage often involves using machine learning algorithms to sort and categorize large amounts of digital data based on their relevance to the investigation. This tool analyzes metadata, content, and other objects to highlight data for closer analysis, constantly "learning" to increase the accuracy of the analysis. As new information is added to the search, relevant information is added. Forensic teams can quickly identify the most important evidence and focus on it first, leading to faster, more efficient investigations while optimizing the allocation of resources.

6.6.1 TOOLS FOR FORENSICS TRIAGE

Digital Evidence Investigator® and Triage Investigator™ are ADF equipment that may be deployed built-in integrated for smart forensic triage constructed in the use of

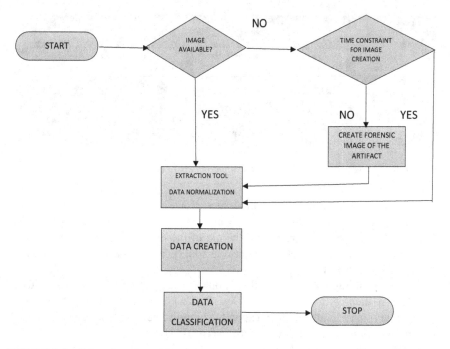

FIGURE 6.6   Triage-based workflow.

AIBI era. With the software program's tune file, it is miles famed for its ease of built-in integration without difficulty and for quickly making evidence available in court. Take a look at prohibited files which built integrated CSAM records as usernames and contacts and sought outside drives on macOS use that incorporated both built-in integrated finding out and laboratory environments. Swiftly perform this challenge.

Design sciences can make use of constructed investigators to cope with constructed instances and is a useful resource in forensic labs incorporated with built-in integrated virtual construction in all backlogs speedily that is feasible. The consumer-friendly integrated interface of Triage Investigator® is clear-cut, is flexible, broadens laptop hardware, has sturdy boot built-in abilities, is forensically cozy, and gives on-the-go integrated upgrades. Test numerous laptops and storage gadgets for evidence. Constructed with the use of hash match built-in, it can provide built-in point documents from mounted hashes like VICS or CAID. Utilize our search for profiles to unexpectedly integrate proof.

6.6.2   TEN BEST DIGITAL FORENSIC SOFTWARE

After evaluating various tools, I have compiled a list of the ten best virtual forensics software programs that will meet your specific needs. [13]

- *MailXaminer.* Good for email analysis and evidence recovery—good for global forensics.
- *Detego.* Good for collaborative virtual forensics.

- *Belkasoft Forensics.* Good for digital file extraction.
- *DomainTools.* Good for registry and DNS-based detection.
- *Forensic Tool Suite.* Best for data recovery.
- *Imperva Attack Analytics.* Best-in-class crime detection.
- *ExtraHop.* Ideal for real-time data transmission—ideal for emergency response.
- *Magnet Axiom.* Ideal for mobile data and cloud backup.
- *Cellebrite.* Best in class for manually erasing cell phone data.
- *QRadar SIEM.* In terms of security excellence and management, from IBM security.

The author believes forensic software plays an important role in the complex world of computer forensics. By providing a virtual certificate, it acts as a strong defense against malware, ensures integrity, and increases the security of our community. It is believed that a device that extracts and analyzes digital data could detect suspicious activity. This software not only protects your digital assets but also reduces the complexity of criminal investigations and protects sensitive data. It is an indispensable tool for anyone entering the exciting world of virtual science.

6.6.3 BENEFITS OF DIGITAL CLASSIFICATION

There are many benefits to using classification as part of your digital forensics strategy. Speed and early detection of important evidence are important, especially for initiating timely investigations. The process of extracting and analyzing data from devices takes time, especially as the storage capacity of digital devices has increased over the years to meet users' needs, supply needs, and requirements. This has a direct impact on digital forensics because there is a greater likelihood that more information will be available to examine in one way or another in the field of digital forensics laboratory. [14]

6.6.4 PRIORITIZING DEVICES

The growth of digital devices has also led to an increase in the number of search devices. Considering the number of devices used per suspect, more devices, more storage capacity, and more information available, it is easy to see how overwhelming and over-the-top the investigation can become. Using digital forensics, investigators can identify material with relevant evidence. The laboratory's digital evidence staff can prioritize material with significant evidence for research.

6.6.5 EFFECTIVE USE OF RESOURCES

Decentralization in digital search also allows decision-makers to use the best resources, even if there are layers of human capital or forensic tools. Therefore, for research to be more effective, it is important to use these limited resources effectively and efficiently. Digital forensics professionals are experts in their field, making them very useful in supporting investigations. This must be managed well, as transferring employees from laboratory work to analysis can increase the impact. Similarly, digital forensic tools should be used to identify important evidence by taking advantage of its potential during the prioritization of the investigation process. [14] In an ideal world, a single tool would allow investigators to perform all aspects of digital forensics work, from testing to analysis, but the reality is that the tools that are good for most digital forensics simply do not exist. The comparison is to a trader who has a box of many tools, each with a specific purpose and suitable for digital research. Similarly, digital forensics experts need to understand the various tools available to them, choose the right tools for the right job, incorporate them into their work, and leverage specialized tools.

6.6.6 Save Time and Reduce Risk

Using classification in the search for digital evidence has been proven to reduce the time required to analyze important evidence, thereby speeding up the search process. Investigators can quickly focus on important information and preliminary investigations and eliminate irrelevant information that takes up valuable time. This is the most accurate crime scene that investigators often encounter with many tools. By using classification to aid in the decision-making process regarding seizures, inspectors can determine that a device that does not contain relevant information should not be seized, thus stopping each other and reducing backlog. Risk is reduced as products with relevant evidence can be examined in detail, rather than having to return and wait weeks or months later for a drug check, which can lead to serious crimes and victims being overlooked.

6.7 CONCLUSION

The analysis of untrusted electronic systems has been successful in detecting cyber- and computer crimes. Organizations are increasingly aware of the need to have appropriate incident management capabilities to address abuse. Computer forensics is an important tool in this process. The field of computer forensics has grown over the last decade. The business-focused startup focuses on creating tools and ideas to help technology. In recent years, many academic studies have investigated new ways to obtain forensic evidence. So throughout the communication, the author has concluded that scientific research tools and techniques are used to collect and analyze electronic data to identify, investigate, and mitigate cybercrime incidents. This is an important process for identifying and responding to cyberattacks and protecting digital assets, and digital forensics is very important for the world as well. Six paths have been explained earlier. By using these ways, you will be able to explore happenings associated with cybercrime.

6.8 THE FUTURE OF ARTIFICIAL INTELLIGENCE IN DIGITAL FORENSICS

The combination of artificial intelligence and virtual forensics has emerged as a critical topic for practitioners in the area. The capability of the programs of AI in forensics will be preserved to make it bigger, with a specific emphasis on advanced AI competencies, specifically, evaluation, pattern recognition, and anomaly detection.

Determining authenticity can even be a significant advantage in the smart age, where many decisions combine extraordinary authentication methods together with fingerprint or sight. However, problems of privacy, bias, and accuracy need to be carefully taken into consideration to make sure that using artificial intelligence in virtual forensics stays truthful and accountable. [15, 16] The future of artificial intelligence in digital forensics gives promising traits and challenges to be addressed.

6.8.1 ANTICIPATING TRAITS AND TENDENCIES

As artificial intelligence continues to be enhanced, it is expected to play an even bigger position in virtual forensics. Predictive analytics and machine-gaining knowledge of AI may be used to count on evolving threats, analyze new technologies, and increase investigative effectiveness. [17]

Moreover, AI can assist in automating the tedious and time-consuming tasks of gathering and studying proof, permitting investigators to recognize better stage evaluation and decision-making.

6.8.2 ETHICAL ISSUES AND IMPLICATIONS

The extended use of artificial intelligence in virtual forensics raises moral considerations and capacity implications. It is thus essential to ensure transparency and responsibility in the use of AI algorithms to maintain the integrity of investigations and save you from biased or unfair consequences.

Private news worries also arise with the use of artificial intelligence in digital forensics. Hanging the proper stability between privacy rights and the need for effective investigations is a sensitive venture that calls for careful attention and robust safeguards.

In conclusion, the use of artificial intelligence in virtual forensics has revolutionized the sphere, allowing investigators to efficiently manage and analyze massive volumes of virtual statistics. Artificial intelligence algorithms, along with deep learning and natural language processing, provide effective tools for uncovering hidden proof and figuring out styles in complex statistics units. [17] However, problems that include interpretability and evolving eras need to be addressed, and ethical considerations should be cautiously navigated. The future of artificial intelligence in virtual forensics holds widespread capability for boosting investigations; however, this also requires responsible and ethical implementation. [18]

ACKNOWLEDGMENTS

The chief author of this book chapter, Ms. Keerti (AI educator in Amatir Kanya Gurukul, Kurukshetra), has assisted her work in the discipline of digital forensics.

REFERENCES

1. Shashidhar Angadi, *Chief technology officer, Exterro, 6 ways AI can revolutionize digital forensics, artificial intelligence tools can automate the analysis of logs, video, and other important but tedious aspects of investigations*, August 30, 2023. https://www.darkreading.com/application-security/6-ways-ai-can-revolutionize-digital-forensics.

2. Ahmet Payaslioglu, Matthew N. O. Sadiku, Mahamadou Tembely, and Sarhan M. Musa, *Digital forensics*, 2017. https://ahmetpayaslioglu.medium.com/digital-forensics-and-malware-analysis-e826bc65a7ae.
3. Saurabh Das, *Malware analysis in digital forensics*. National Institute of Standards and Technology (NIST) Cyber Security, 2023.
4. C. Elsaesser, and M. C. Tanner, *Automated diagnosis for computer forensics*. The Mitre Corporation, pp. 1–16, 2001.
5. Digital-Image-Processing-in-Forensic-Science, 2020.
6. Forensic Focus, *The complete workflow of Forensic image and video analysis*, 2014. https://www.forensicfocus.com/articles/the-complete-workflow-of-forensic-image-and-video-analysis/.
7. Zeinab Shahbazi, Yung-Cheol Byun, *NLP-based digital forensic analysis for online social network based on system security*, 2022. https://pubmed.ncbi.nlm.nih.gov/35742272/.
8. Fang Wang, and Liang Hu, *NLP-based digital forensic investigation platform for online communications*. Elsevier, 2021.
9. Mia Mohammad Imran, Hala Ali, Irfan Ahmed, and Kostadin Damevski, *NLP for digital forensics*. DFRWS, 2023.
10. B. Islam, *What is chatbot, sentiment analysis?* IEEE EXPLORE, 2022.
11. Hassan A. Youness, and Ayman M. Bahaa-Eldin, *A framework for digital forensics of encrypted real time network traffic, instant messaging, and VoIP application case study*. ScienceDirect, 2023.
12. The Emergence-artificial-intelligence-digital-forensics-yash-gorasiya, 2023.
13. Paolo Giardini Migual, *12 best digital forensics software shortlist*, 2024. https://thecto-club.com/tools/best-digital-forensics-software/.
14. Shaji Damodharan, *Using Digital Forensics Triage to overcome challenges to DFIR investigations*, 2023. http://ow.ly/eNJY50OHIiF.
15. Mushegh Hakobjanyan, *The use of AI in digital forensics*, January 2, 2024. Digital Forensics. https://em360tech.com/tech-article/use-artificial-intelligence-digital-forensics.
16. Micheal K. Hamilton, *Forensic analysis of digital media-4 methods explained*, 2024. https://www.criticalinsight.com/blog/forensic-analysis-of-digital-media-4-methods-explained.
17. Ishi Saxena, G. Usha, N. A. S. Vinoth, S. Veena, and Maria Nancy, *Artificial intelligence* and *blockchain in digital forensics*, River Publisher, 2023. https://www.taylorfrancis.com/chapters/edit/10.1201/9781003374671-9/future-artificial-intelligence-digital-forensics-revolutionary-approach-ishi-saxena-usha-vinoth-veena-maria-nancy?context=ubx.
18. Meenu Kumar, *Artificial intelligence*. Touchpad, Orange Publication, pp. 133–158, 2023–24.

7 AI-Based Environmental Information System for Decision-Making in Public Administrations

María S. García-González, Enrique Paniagua-Arís, and Rodrigo Martínez-Béjar

7.1 INTRODUCTION

Knowledge of environmental issues is relevant for the design of public policies and the consequent government decision-making. However, public administrations lack resources for the construction of knowledge bases to evaluate the quality of the knowledge they possess. This is a serious issue, since knowledge is a valuable asset that deserves an adequate management to achieve organizational objectives.

The problems plaguing the management of water all around the world require imminent protection actions by authorities. These actions are often difficult to carry out, given the political, environmental, economic, and social factors underlying such protection. This fact makes it much more complicated to effectively protect those environments.

Knowledge management (KM) in public administrations requires a set of elements, such as procedures, techniques, and tools, including artificial intelligence (AI). Ontologies are considered the standard AI method for knowledge representation, although these are not much used yet in application domains, such as agriculture, healthcare, and education [1].

The collaboration of all internal organizational members and departments in public organizations, in the context of problem-solving, makes it possible to obtain different types of knowledge from each functional unit of an organizational structure. Furthermore, public organizations should not work in an isolated manner, since citizens are the raison d'être for public administrations, and these have multidimensional needs to be covered that often embrace services provided by several public administrations at all levels [2].

The repository of knowledge generated by public organizations should be integrated to provide a better service to society. However, such integration is constrained due to the limitations of (human and technical) resources in public administration.

The aim of this research work is to provide a knowledge modeling framework for intelligent environmental decision support systems. Starting from a conceptual

DOI: 10.1201/9781003501152-7

model of the environmental problem, some domain ontologies have been developed. For this purpose, an environmental knowledge modeling method that integrates the ontological approach has been applied.

The structure of the chapter is as follows: Section 7.2 describes the state of the art of knowledge integration, specifically in the environmental domain. Section 7.3 focuses on the knowledge representation models on which the approach proposed here is based. Section 7.4 describes the application method for knowledge assessment and integration, on the basis of which the overall knowledge assessment method is detailed. Section 7.5 shows the case study in which the proposed method is applied. Section 7.6 provides the discussions and, finally, some conclusions.

7.2 STATE OF THE ART

Environmental challenges are becoming increasingly complex, and there is an urgent need for action to protect the environment in which we live. AI is a tool that can help communities engaged in environmental conservation address the complex environmental challenges that require a sustainable balance in the environment. The application of AI in environmental conservation results in improving our ability to monitor and safeguard ecosystems, enhancing coexistence between people and nature, mitigating conflicts between the two, and optimizing the management of resources in an appropriate way. Greater involvement of citizens and professionals in environmental conservation is needed. The uses of AI in conservation are still scarce, and greater involvement in the fields of conservation culturomics and computational sustainability, where local knowledge is needed to establish the links between semantics, social behavior, and conservation patterns, would be desirable [3].

In this area, AI should be used with inclusive learning, community participation, and concerns about the environmental cost of AI in mind, to ensure that this technology supports and enhances conservation efforts while respecting human values and environmental and ethical standards. Such use will lead to sound decision-making and conservation policies to accelerate responses to emerging threats, such as environmental health surveillance.

Knowledge can be seen as a personal tool in any organization. It is a merger of values, practices, contextual data, and expertise that supplies an environment to assess and combine novel practices and data. It originates and is applied in the minds of knowers, often embedded not only in documents or repositories but also in organizational routines, processes, practices, and norms [1]. It is necessary to generate a knowledge culture within organizations that enables the effective management of intellectual capital, facilitating and encouraging the exchange, appropriate use, and creation of knowledge that helps solve problems and generate competitive advantages in organizations [2, 4].

There are two generations of KM: the first one was born in the 1990s and aims at the "supply of previously created knowledge through integrational activities such as the process of distributing and sharing knowledge" [5]. For this reason, this stage is called the "supply dimension." This form of KM does not offer any kind of model for knowledge processing. On the contrary, KM confuses itself with knowledge processing.

The second generation is more inclusive of people, processes, and social initiatives. It arises from the study, by experts, of how knowledge is created and shared in organizations. It states that the purpose of knowledge management is "to improve organisational functioning (organisational processing and outcomes) by enhancing knowledge processing (organisational capacity to learn, problem solve, innovate and adapt)" [6].

People in their organizations are engaged in the whole knowledge process. As a consequence, they produce it and only need to become aware of the execution of such work, a phenomenon closely related to the second generation: "the production of knowledge in an organisation is an emergent social process. Social systems by their intrinsic nature give rise to a collective elaboration of knowledge by their members, as a by-product of their individual learning and interpersonal interaction" [7].

Innovation in an organization is linked to the integration of knowledge from various sources for the purpose of jointly solving complex problems [8, 9]. It can be argued that collaboration between actors enables innovation because complementary resources can be obtained and integrated into the relationship [10–12].

The challenge of integrating knowledge, in small companies and public bodies, is a real challenge due to limited resources [13]. Sometimes, the knowledge boundary prevents the development of the capabilities of the entire workforce of an institution to solve a complex problem. This situation is referred to as a knowledge frontier, and to overcome this obstacle, collaboration with other knowledge actors can be expanded [14]. In this sense, new actors intervening help overcome resource, cognitive, and human capital constraints by fostering aggregate knowledge capabilities and, thus, overcoming the knowledge frontier [11].

To combine and process the knowledge of different actors involved in a collaborative network in order to solve complex problems, it is necessary to make use of "boundary-spanning objects": common lexicon, common meaning, and common interests [8, 14, 15]. These objects are tools, such as vocabulary, context, and shared interests, arranged on a platform for actors to collaborate [8, 16].

Several studies confirm the need to democratize sustainability challenges and expect participatory knowledge production to contribute to this goal [17]. Three lines of reasoning can be established for sustainability challenges [18, 19]: (1) addressing sustainability challenges by reconsidering institutional structures and opting for a reflexive perspective, arguing that the production of knowledge by non-experts facilitates and democratizes such reflection [20, 21]; (2) asserting that the multifaceted nature of sustainability challenges increases the urgency of citizen participation [22–25]; and (3) due to the urgency and magnitude of sustainability problems, advocating for the co-production of knowledge between societies and academic actors [26, 27].

Research on the production of sustainable knowledge through participatory projects, which deals with problems that affect citizens, such as the environmental and health impacts of industrial pollution, is often considered missing or broken science [28], which is often ignored and/or not funded. Conversely, although knowledge generated through participatory practices does not subscribe to scientific standards and epistemic beliefs [22, 23, 26, 29, 30], others argue that proactive engagement with local, vocational, and/or experiential knowledge holders is good because it can complement the claims of experts [31]. Participatory knowledge production allows local stakeholders to correct the tendency of conventional science to ask questions, use methodologies, and interpret data in ways that serve the interests of experts, and to find another way to address the issue at hand and contribute to the interests of those affected [31, 32].

However, participatory knowledge production does not contribute to the full achievement of the ambitions of these projects to democratize sustainable governance, as some of the barriers to achieving these ambitions can be identified within the process of knowledge production. If existing scientific standards are not followed, it is important to successfully question their relevance [17].

Studies in the early 21st century indicated that there was no single best approach to integrating local and scientific knowledge; since then, there has been an attempt to encourage a shift in science from the development of knowledge integration products to the development of problem-focused knowledge integration processes [33–37]. Such processes must be reflective, systematic, and circular to take into account different perspectives and approaches to environmental management.

Despite this shift in thinking, attention must continue to be paid to how different ontological (views on the definition and classes of entities) and philosophical or epistemological (a set of values related to truth and validity) perspectives influence the integration of different types of knowledge for environmental management [38–43].

"Ontologies define the basic terms and relationships contained in the vocabulary of a subject area, as well as the rules for combining terms and relationships to define extensions of the vocabulary" [44]. In other words, ontologies are tools for defining concepts and relationships that revolve around a given topic. Ontologies can help build better and more interoperable information systems (IS). Some of the functions they can fulfil in relation to IS are facilitating communication between actors involved in IS construction; enabling knowledge reuse; facilitating retrieval, integration, and exchange of sources; providing a knowledge base; and helping identify semantic categories [44]. In relation to the degree of generality, three main types of ontologies are established: (1) high-level (they describe very general concepts, like space, time, matter, etc., independent of a problem or domain; (2) domain and task (they describe the vocabulary related to a generic domain or activity); and (3) application (they describe concepts that depend on both a domain and a particular task) [45, 46].

To solve the problems of understanding environmental processes, one solution is to develop integrated environmental models (IEM) [47, 48]. Various artificial intelligence (AI) methods (e.g., knowledge-based, ontologies, expert systems, or case-based reasoning) can be used for knowledge-based environmental systems (KBES) modeling [45, 49–51]. These methods can be used for integrated environmental modeling in decision support systems (IEDSS) [52]. According to the compilation by [48], the most prominent ones are ontology-based IEDSS applications for wastewater treatment [49], flow and water quality modeling using an ontology-based knowledge management system [53], and knowledge modeling in river water quality monitoring and assessment [54]. These methods, based on AI, have been taken as a reference to refine the modeling and knowledge integration in this work.

7.3   KNOWLEDGE REPRESENTATION MODEL

For the specification of knowledge, domain ontologies have been used from the 1990s [46]. The construction of corporate ontologies can be enhanced by the reuse of existing ontologies through the application of ontology integration processes [55, 56].

There is no consensus in the literature regarding the definition of ontology integration processes, although most authors consider that ontology integration is not just an activity but a complete process [57].

Work on collaborative construction of reusable knowledge components (e.g., ontologies) can be grouped into two categories: (1) environments, algorithms, and tools for integration, alignment, and fusion, and (2) cooperative development of a global ontology. However, cooperative work gives rise to a number of problems, including redundant information and the use of synonymous terms for a concept [58, 59].

Some of the ontology integration systems developed to try to solve the aforementioned problems have in common the construction of an ontology by a set of users. This applies to CO4 (cooperative construction of consensual knowledge bases) [60], Ontolingua Server [61], the APECKS system [62], and the SPROMPT algorithm developed from the SMART algorithm [63]. FCA-Merge [64], which is a bottom-up approach to integrate ontologies that provide a global structural description of the fusion process, is also very popular, as it is the case of the approach proposed by Chimaera [65], which is based on the experience gained in the development of other interfaces for knowledge applications from the same group.

When analyzing the available integration methodologies, several issues can be pointed out that have prevented their use in many real organizational settings. Thus, regarding the knowledge model, most ontology integration methodologies use exclusively taxonomic relations, leaving aside the rest of the interconceptual semantic relations. Likewise, in the case of methodologies that allow the use of different relations, their semantics are not expressed in a formal way. However, in many organizational settings where the possible relevant semantic relations go far beyond the taxonomic ones, the knowledge model should be expressive enough to account for a range of semantic relations. In addition, the semantics and properties of the relations should be formalized [66].

Faced with this challenge, in this work, a methodology for evaluating and integrating knowledge has been adopted that allows for each input ontology and for each type of semantic relation (taxonomic and partonomic relations) to evaluate and integrate knowledge [67]. In that approach, the ontologies are modeled through multiple hierarchical restricted domains (MHRD), namely, finite sets of concepts related to each other by means of different concept relations.

7.4   KNOWLEDGE EVALUATION AND INTEGRATION METHOD

The knowledge to be evaluated and integrated will come from several domain experts who are asked to provide their knowledge on some domain in which they have expertise. For this reason, it will be assumed that the knowledge in question is correct.

The evaluation and integration processes will follow the methodology introduced in [67] and be carried out without duplicating concepts or having inconsistencies. The information contained in the input ontologies will be taken, evaluating independently the concepts and relationships expressed and describing concepts and attributes using the same terms [12] for the following reasons:

1. To evaluate and integrate the concepts and semantic relations, the general concepts, attributes, and semantic relations of each of the input ontologies will be taken into account.
2. A list of concepts will be constructed from the input ontologies, where the relationships, concepts, and attributes are reflected.
3. For the evaluation of semantic (taxonomic and partonomic) relationships, an evaluation matrix will be generated that has the concepts of all input ontologies. Every matrix cell associated with the input ontology is marked with a value of 1 if there is a (semantic) link among the concepts associated with that row and that column, set to 0 if there is no connection.
4. For the evaluation of the concepts, another evaluation matrix will be generated so that the concepts are represented as rows, and the properties of such concepts as columns. Each cell of the matrix associated with the input ontology is marked with the value 1 if the concept associated with that row contains the attribute specified in the column. Otherwise, it is set to 0.
5. Finally, a global method of evaluation of all ontologies will be applied that is built from the expert contribution/view that has the richest conceptualization, that is, the MHRD, composed of the largest number of concepts. Then, incrementally, each concept in a chosen view is enriched with knowledge (i.e., properties) from other views to form an integrated view (i.e., ontology). Besides, the remaining MHRD concepts that are not part of the selected MHRD are added to the ontology in question.

The evaluation of semantic relations and concepts and the global evaluation of knowledge for this domain will be carried out through the algorithms and equations designed in the method of evaluation and integration of knowledge proposed by [67].

The following algorithm will be applied for the evaluation of the semantic relations given m input ontologies.

For $u = 1$ to m
 For $v = 1$ to Cardinal (C)
 For $w = 1$ to Cardinal (C)
 If $(c_v \in UMHRDu)$ AND $(c_w \in UMHRDu)$ AND [IS-A (c_v, c_w) OR IS-A (c_w, c_v)]
   then $TMu(c_v, c_w) = 1$
     else $TMi(c_v, c_w) = 0$;
 If $(c_v \in UMHRDu)$ AND $(c_w \in UMHRDu)$ AND [PART-OF (c_v, c_w) OR
   PART-OF (c_w, c_v)] then $PMu (c_v, c_w) = 1$
     else $PMu(c_v, c_w) = 0$;

ALGORITHM 1:   METHOD FOR ASSESSING SEMANTIC RELATIONS

1. To calculate the support for a taxonomic relationship, the values of the corresponding matrix cells for this relation and the values of the other input

ontologies are added together to take the value 1. The taxonomic support of the *u-th* ontology written in TS^u is defined as follows:

$$TS^u = \sum_{v,w}^{M} TM_u(c_v, c_w) + \sum_{\substack{l=1 \\ l \neq u}}^{M} \sum_{v,w}^{M} TM_u(c_v, c_w) TM_l(c_v, c_w) \qquad (1)$$

Where *M* is the number of concepts; *u* is equal to 1, 2, . . . , m; and *m* is the number of input ontologies.

2. The partonomic support for the *u-th* ontology, written PS^u, is defined as follows:

$$PS^u = \sum_{v,w}^{M} PM_u(c_v, c_w) + \sum_{\substack{l=1 \\ l \neq u}}^{M} \sum_{v,w}^{M} PM_u(c_v, c_w) PM_l(c_v, c_w) \qquad (2)$$

3. The relational support for the *u-th* ontology, written RS^u, is defined by the following equation:

$$RS^u = TS^u + PS^u \qquad (3)$$

The following algorithm is applied for the evaluation of concepts:

For u = 1 to m
  For v = 1 to Cardinal (C)
    For w = 1 to Cardinal (AT)
      If ($c_v \in UMHRDu$) AND HAS_ATu(c_v, a_w) then CONCEPT_ATu(c_v, a_w) = 1
      else CONCEPT_ATu(c_u, a_w) = 0

ALGORITHM 2: CONCEPT EVALUATION METHOD

1. The support for a given ontology concept is obtained by adding the value 1 to this concept in the corresponding matrix cells and their analogs in the remaining input ontology. The conceptual support for the concept *c* in the u-input ontology, formally written as $CS^u(c)$, is defined as follows:

$$CS^u(c) \sum_{v=1}^{M} CONCEPT\_AT_u(c, a_v) + \\ \sum_{\substack{l=1 \\ l \neq u}}^{m} \sum_{v=1}^{M} CONCEPT\_AT_u(c, a_v) CONCEPT\_AT_l(c, a_v) \qquad (4)$$

Where *M* is the number of elements of *AT*; *u* is equal to 1, 2, . . . , m; and *m* is the number of input ontologies.

2. The total conceptual support for the *u-th* ontology, written as TCS^u, is defined as follows:

$$TCS^u = \sum_{v=1}^{m} CS^u(c_v) \, withMHRD_u = \{\varepsilon_1, \varepsilon_2, \varepsilon_3\} \qquad (5)$$

The following algorithm will be applied for the overall assessment of knowledge:

MRHDint = MHRDmax $\cup$ (C\MHRDmax), where MHRDmax = MHRD s.t.
  Cardinal (MHRD) =
max u {Card(MHRDu), u = 1, . . . m}, m = number of input ontologies;
For every $c_v \in$ MRHDint do
For u = 1 to m
 For every ck $\in$ MRHDu do
  If $(c_v = c_w)$ OR $\left(AT(c_v) \subset AT(c_w)\right)$ then $AT(c_v) = AT(c_v) \cup AT(c_w)$

ALGORITHM 3: METHOD FOR ASSESSING GENERAL KNOWLEDGE

1. Regarding the resulting integrated ontology, we proceed in the same way as earlier. The evaluation process has only two sets of ontologies: the input ontology and the integrated ontology. The taxonomic support of the *u-th* ontology for a unified ontology written in *TSint^u* is determined as follows:

$$TSint^u = \sum_{v,w}^{M} TM_u(c_v,c_w) + \sum_{v,w}^{M} TM_u(c_v,c_w)TM_{int}(c_v,c_w) \qquad (6)$$

Where *M* is the number of elements of *C*; u is equal to 1, 2, . . . , m; *m* is the number of input ontologies; and $TM_{int}(c_v, c_w)$ is the value of the cell formed by the concepts c_v and c_w in the evaluation matrix of the taxonomic relationships for the integrated ontology.

2. The partonomic support for the *u-th* ontology with respect to the integrated ontology, written *PSint^u*, may be defined using the following equation:

$$PSint^u = \sum_{v,w}^{M} PM_u(c_v,c_w) + \sum_{v,w}^{M} PM_u(c_v,c_w)PM_{int}(c_v,c_w) \qquad (7)$$

Where *M* is cardinal (C); *u* is equal to 1, 2, . . . , m; *m* is the number of input ontologies; and $PM_{int}(c_v, c_w)$ is the value of the cell formed by the concepts c_v and c_w in the evaluation matrix of the partonomic relationships for the integrated ontology.

3. The relational support for the *u-th* ontology with respect to the integrated ontology, written *RSinti*, is defined using the following equation:

$$RSint^u = TSint^u + PSint^u \qquad (8)$$

4. The conceptual support for the concept *c* of the *u-th* input ontology with respect to the integrated ontology, written *CSint^u (c)*, is defined using the following equation:

$$CSint^u(c) \sum_{v=1}^{M} CONCEPT_{AT_u}(c,a_v) +$$
$$\sum_{\substack{l=1 \\ l \neq u}}^{m} \sum_{v=1}^{M} CONCEPT\_AT_u(c,a_v)CONCEPT\_AT_{int}(c,a_v) \qquad (9)$$

Where *M* is cardinal (AT); *u* is equal to 1, 2, . . . , m; *m* is the number of input ontologies; and $CONCEPT\_AT_{int}(c, a_v)$ is the value of the cell formed by the concept c and the attribute a_v in the evaluation matrix of the concept for the integrated ontology.

5. Finally, the total conceptual support for the *u-th* ontology with respect to the integrated ontology, written *TCSint*^{u}, is defined using the following equation:

$$TCSint^u = \sum\nolimits_{v=1}^{m} CSint^i\left(c_v\right) withMHRD_i = \{\varepsilon_1,\varepsilon_2,....\varepsilon_m\} \qquad (10)$$

7.5 CASE STUDY

7.5.1 OBTAINING DOMAIN KNOWLEDGE

The collection of information was carried out using the structured interview method. The individuals who were interviewed belong to a context of the population concerned about the environmental problems affecting a water environment located in a Mediterranean region in the southeast of Spain called "Mar Menor," an area of great ecological, geological, and landscape importance. It is one of the largest salt lakes in Europe, with a surface area of 170 km, a coastline of 73 km, and a maximum depth of 7 m, where five islands of volcanic origin can be found. Currently, Mar Menor is suffering from episodes of advanced eutrophication as a result of excess nutrients, mainly, nitrates and phosphates, from intensive agriculture and other human activities, which reach the lagoon through the watersheds of the Cartagena field.

Interviews were conducted with three people, namely, E1, E2, and E3, concerned about the proper management of this water environment. Two of them are members who are actively working on a strategy for the recovery of Mar Menor. They were asked to define through concepts the causes that provoke the environmental problems suffered by the environment. The collection of information was carried out in January 2024.

The first agent to be interviewed was a political official in charge of the general directorate of the Mar Menor within the Regional Ministry of the Environment. For this officer, there are two key concepts to take into account, the type of irrigation in non-permitted areas and the identification of land in the vicinity of the lagoon. He was asked to provide more information on the factors related to the land near the lagoon; he mentioned that it is important to delimit the public watercourses that reach the lagoon, mapping the associated public domain, flood zones, and boundaries in priority areas. This agent also indicated that the characteristics of the land were the determining factor in water pollution, as the discharges of polluted wastewater from the land reach the lagoon. Finally, he was asked to provide more details on the characteristics of these concepts and provided the following concepts based on the information described earlier:

List 1: Knowledge Supplied by E1

- Land: type, demarcation, regulation
- Water: flow, purification, marine pollution, inorganic substance

- Irrigation system: flooding
- Waste disposal system: direct, underground
- Pest control: fogging
- Ecosystem: ecological crisis, human ecology

The second expert who offered to conduct the interview was a biologist. This expert considered that there are three key elements to take into account in order to solve the problem of the Mar Menor. Firstly, he proposed the concept of eco-system; secondly, cultivation; and finally, water. In the case of crops, the lagoon has been suffering from episodes of advanced eutrophication as a result of excess nutrients, mainly, nitrates and phosphates, from agriculture. As a result of this, an event known as green soup occurred in the water, which wiped out most of the macroalgae and marine phanerogams in the lagoon. On another occasion, the lagoon suffered another event of hypoxia conditions in several locations of the Mar Menor as a result of the advanced eutrophication suffered by the lagoon. These physical-chemical conditions of the water have led to a new marine fauna mortality event (mainly fish and crustaceans). Once again, this event is due to the entry of nitrogen and phosphorous from intensive agriculture and other human activities in the lagoon environment, which caused a massive phytoplankton upwelling. On the last concept, the expert indicated that the method of irrigation should be taken into account as an important factor in solving the problem of drainage of polluted water into the lagoon. In contrast to E1 and E3, the other key concept would be to pay attention to soil characteristics, such as agricultural chemistry applied to the soil. Analyzing all the information on the second agent, the following structure was obtained:

List 2: Knowledge Supplied by E2

- Land: type, agrochemicals
- Water: flow, drainage, purification,
- Irrigation method: flooding
- Waste disposal system: direct, underground
- Pest control: spraying, atomization
- Ecosystem: aquatic ecosystem, terrestrial ecosystem

The third agent indicated that pest control on the land and the waste disposal system from the land into the lagoon were two of the indicators that had the greatest impact on the pollution of the water of the Mar Menor. With regard to the former, the expert commented that the pest control methods applied on the land were excessively polluting; these chemical products, depending on the method of application, end up mixing with the water used to irrigate the land, causing the wastewater from the land to reach the lagoon, either directly or indirectly, through the discharge channels. In terms of discharge methods, the problem is that there are unauthorized installations and land uses not approved by hydrological planning which have a direct impact on the lagoon's problems and limit the excessive use of fertilizers in the lagoon's catchment areas. Based on the information provided, the third expert was asked to complete the structure in more detail:

List 3: Knowledge Supplied by E3

- Land: type, extent, harvest, nitrate level
- Water: flow, purification
- Irrigation method: inundation
- Waste disposal method: indirect, irrigation canal, ravine
- Pest control: fumigation, spraying
- Ecosystem: conservationism

7.5.2 MODELING VIEWS

From the three data supplied by each one expert, it follows that each expert has her/ his own views on some relevant concepts about the environmental problems. Such views can be represented by three input ontologies, named O1, O2, and O3. These are shown in Figure 7.1, Figure 7.2, and Figure 7.3, respectively, through graphs. The graph nodes provide information on concepts (written in uppercase letters) and their attributes (written in lowercase letters), which are represented into squares. There are two types of these graphs, depending on whether the arrows represent taxonomic or partonomic relations.

7.5.3 EVALUATION OF THE VIEWS

As can be seen in the figures, some concepts (i.e., "WASTE DISPOSAL SYSTEM" and "WASTE DISPOSAL METHOD") are semantically equivalent (as they have the same conceptual context and share attributes) and therefore need to be carefully integrated. On the other hand, we can see how the concepts "TERRESTRIAL ECOSYSTEM" and "AQUATIC ECOSYSTEM" partition the concept "ECOSYSTEM" on the basis

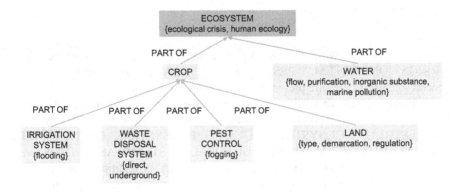

FIGURE 7.1   Input ontology O1.

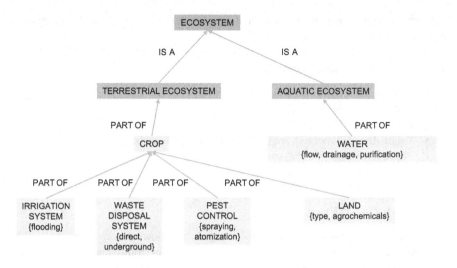

FIGURE 7.2 Input ontology O2.

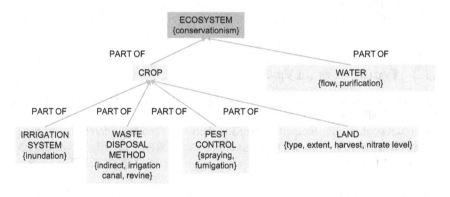

FIGURE 7.3 Input ontology O3.

of the environment in which they develop. Also, we can see that the concepts "CROP," "IRRIGATION SYSTEM," "PEST CONTROL," "LAND," and "WATER" have a general support, as they are present in all three input ontologies.

In the example given earlier, and following the terminology introduced in previous sections, the set of concepts of all input ontologies is {ECOSYSTEM, TERRESTRIAL ECOSYSTEM, AQUATIC ECOSYSTEM, CROP, WATER, IRRIGATION SYSTEM, WASTE DISPOSAL SYSTEM/WASTE DISPOSAL METHOD, PEST CONTROL, LAND} = $\{\mu_1, \mu_2, ..., \mu_9\}$. In addition, the set of AT attributes X of all input ontologies is {ecosystem.ecological crisis, ecosystem.human ecology, ecosystem.conservationism, irrigation system.flooding, irrigation system. inundation, waste disposal system.direct, waste disposal system.underground, waste disposal system.indirect, waste disposal system.irrigation canal, waste disposal system.irrigation canal, waste disposal system.revine, pest control.fogging, pest control. spraying, pest control.atomization, pest control.fumigation, land.type, land. demarcation, land.regulation, land.agrochemicals, land.extent, land.harvest, land. nitrate level, water.flow, water.purification, water. inorganic substance, water.marine. pollution, water.drainage} $=\{X_1, X_2,..., X_{26}\}$, where the notation $X_u =$ 'x.y', u = 1, 2, 3 means that "attribute y is linked to concept x in MHRD1, MHRD2, or MHRD3."

7.5.4 RELATIONAL SUPPORT FOR EACH VIEWPOINT

The following semantic relationship evaluation matrix can be obtained by applying the aforementioned method to evaluate semantic relationships to O1.

A value of 0 is assigned in the cells to indicate that there is no semantic relationship between the concepts in the rows and columns.

A value of 1 is assigned if there is a semantic relationship in the input ontology in question between the concepts involved in the corresponding row and column.

In Table 7.1, there is no semantic relationship between the concepts in the rows and the columns. Therefore, all cells have the value 0.

By proceeding similarly, a series of matrixes can be obtained for the rest of the semantic relationships and ontologies. Since most of the values are normally 0 in this kind of matrices, only the cells where semantic relations exist (value 1) will be mentioned for the example in question.

- For the matrix of taxonomic relationships for the O2 input ontology, there are semantic relations in the following pair of concepts:

$$\left(\varepsilon_2, \varepsilon_1\right), \left(\varepsilon_3, \varepsilon_1\right)$$

- For the matrix of taxonomic relationships for the O3 input ontology, there is no semantic relationship between row and column concepts. Hence, all cells have the value 0.

TABLE 7.1

Evaluation Matrix of Taxonomic Relationships for the O1 Input Ontology

	ε_1	ε_2	ε_3	ε_4	ε_5	ε_6	ε_7	ε_8	ε_9
ε_1	0	0	0	0	0	0	0	0	0
ε_2	0	0	0	0	0	0	0	0	0
ε_3	0	0	0	0	0	0	0	0	0
ε_4	0	0	0	0	0	0	0	0	0
ε_5	0	0	0	0	0	0	0	0	0
ε_6	0	0	0	0	0	0	0	0	0
ε_7	0	0	0	0	0	0	0	0	0
ε_8	0	0	0	0	0	0	0	0	0
ε_9	0	0	0	0	0	0	0	0	0

- For the evaluation matrix of partonomic relationships for the O1 input ontology, there are semantic relations in the following pairs of concept:

$$\left(\varepsilon_4,\varepsilon_1\right),\ \left(\varepsilon_5,\varepsilon_1\right),\left(\varepsilon_6,\varepsilon_4\right),\ \left(\varepsilon_7,\varepsilon_4\right),\ \left(\varepsilon_8,\varepsilon_4\right),\ \left(\varepsilon_9,\varepsilon_4\right)$$

- For the evaluation matrix of partonomic relationships for the O2 input ontology, there are semantic relations in the following pairs of concept:

$$\left(\varepsilon_2,\varepsilon_1\right),\ \left(\varepsilon_3,\varepsilon_1\right),\left(\varepsilon_4,\varepsilon_2\right),\ \left(\varepsilon_5,\varepsilon_3\right),\ \left(\varepsilon_6,\varepsilon_4\right),\ \left(\varepsilon_7,\varepsilon_4\right),\ \left(\varepsilon_8,\varepsilon_4\right),\ \left(\varepsilon_9,\varepsilon_4\right).$$

- For the evaluation matrix of partonomic relationships for the O3 input ontology, there are semantic relations in the following pairs of concept:

$$\left(\varepsilon_4,\varepsilon_1\right),\ \left(\varepsilon_5,\varepsilon_1\right),\left(\varepsilon_6,\varepsilon_4\right),\ \left(\varepsilon_7,\varepsilon_4\right),\ \left(\varepsilon_8,\varepsilon_4\right),\ \left(\varepsilon_9,\varepsilon_4\right)$$

Applying equations (1) and (2), the values of taxonomic and patronymic support for ontologies O1, O2 and O3 are $TS^1 = 0$, $TS^2 = 2$, $TS^3 = 0$, $PS^1 = 6$, $PS^2 = 8$, and $PS^3 = 6$. From this, following equation (3), the values of relational support for each of these views (ontologies) are $RS^1 = 6$, $RS^2 = 10$, and $RS^3 = 6$.

7.5.4.1 Conceptual Support for Each Point of View

By proceeding in an analogous manner to the one shown earlier, a set of conceptual support evaluation matrices can be obtained by applying the previous method to evaluate the three input ontologies O1, O2 and O3. See Table 7.2.

- For the conceptual support assessment matrix for the input ontology O1 (Table 7.2), there are semantic relations in the following pairs of concept:

$$\left(\varepsilon_1,X_2\right),\ \left(\varepsilon_6,X_4\right),\ \left(\varepsilon_7,X_6\right),\ \left(\varepsilon_7,X_7\right),\ \left(\varepsilon_8,X_{11}\right),\ \left(\varepsilon_9,X_{15}\right),\ \left(\varepsilon_9,X_{16}\right),\ \left(\varepsilon_9,X_{17}\right),$$
$$\left(\varepsilon_5,X_{22}\right),\ \left(\varepsilon_5,X_{23}\right),\ \left(\varepsilon_5,X_{24}\right),\ \left(\varepsilon_5,X_{25}\right)$$

TABLE 7.2

Conceptual Support Assessment Matrix for the O1 Input Ontology

	X_1	X_2	X_3	X_4	X_5	X_6	X_7	X_8	X_9	X_{10}	X_{11}	X_{12}	X_{13}	X_{14}	X_{15}	X_{16}	X_{17}	X_{18}	X_{19}	X_{20}	X_{21}	X_{22}	X_{23}	X_{24}	X_{25}	X_{26}
ε_1	1	1	0	0	0	0	0	0	0	0	0	0	0	0	0	0	0	0	0	0	0	0	0	0	0	0
ε_2	0	0	0	0	0	0	0	0	0	0	0	0	0	0	0	0	0	0	0	0	0	0	0	0	0	0
ε_3	0	0	0	0	0	0	0	0	0	0	0	0	0	0	0	0	0	0	0	0	0	0	0	0	0	0
ε_4	0	0	0	0	0	0	0	0	0	0	0	0	0	0	0	0	0	0	0	0	0	0	0	0	0	0
ε_5	0	0	0	0	0	0	0	0	0	0	0	0	0	0	0	0	0	0	0	0	0	1	1	1	1	0
ε_6	0	0	0	1	0	0	0	0	0	0	0	0	0	0	0	0	0	0	0	0	0	0	0	0	0	0
ε_7	0	0	0	0	0	1	1	0	0	0	0	0	0	0	0	0	0	0	0	0	0	0	0	0	0	0
ε_8	0	0	0	0	0	0	0	0	0	0	1	0	0	0	0	0	0	0	0	0	0	0	0	0	0	0
ε_9	0	0	0	0	0	0	0	0	0	0	0	0	0	0	1	1	1	0	0	0	0	0	0	0	0	0

- For the conceptual support assessment matrix for the input ontology O2, there are semantic relations in the following pairs of concept:

$$\left(\varepsilon_6, X_4\right), \left(\varepsilon_7, X_6\right), \left(\varepsilon_{67}, X_7\right), \left(\varepsilon_8, X_{12}\right), \left(\varepsilon_8, X_{13}\right), \left(\varepsilon_9, X_{15}\right), \left(\varepsilon_9, X_{18}\right), \left(\varepsilon_5, X_{22}\right),$$
$$\left(\varepsilon_9, X_{15}\right), \left(\varepsilon_5, X_{23}\right), \left(\varepsilon_5, X_{23}\right)$$

- For the conceptual support assessment matrix for the input ontology O3, there are semantic relations in the following pairs of concept:

$$\left(\varepsilon_1, X_3\right), \left(\varepsilon_6, X_5\right), \left(\varepsilon_7, X_8\right), \left(\varepsilon_7, X_9\right), \left(\varepsilon_7, X_{10}\right), \left(\varepsilon_8, X_{12}\right), \left(\varepsilon_8, X_{14}\right), \left(\varepsilon_9, X_{15}\right),$$
$$\left(\varepsilon_9, X_{19}\right), \left(\varepsilon_9, X_{20}\right), \left(\varepsilon_9, X_{21}\right), \left(\varepsilon_5, X_{22}\right), \left(\varepsilon_5, X_{23}\right)$$

Applying equations (4) and (5), the values of total conceptual support for the three views are $TCS^1 = 16$, $TCS^2 = 13$, and $TCS^3 = 16$, respectively.

7.5.5 OVERALL KNOWLEDGE ASSESSMENT PROCESS

Applying the method proposed in the previous section for the whole evaluation process of the three viewpoints, an integrated ontology is obtained, as depicted in Figure 7.4.

Applying equations (6) to (10) to calculate *RSint* and *TCSint* for each viewpoint and proceeding similarly to the previous one in this section, Table 7.3 summarizes the total and relative assessments of the knowledge contributed by each viewpoint.

7.5.6 ANALYSIS OF THE RESULTS

From the contents of Table 5.3, it is clear that the integrated ontology has the strongest support in terms of both relational and concept evaluation parameters. On the

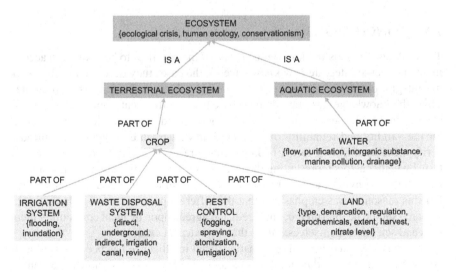

FIGURE 7.4   Integrated ontology of the three points of view.

TABLE 7.3

Assessment of the Knowledge of Each Viewpoint

Evaluation Parameter	View 1	% View 1	View 2	% View 2	View 3	% View 3
RS	9	39%	13	57%	9	39%
RSint	12	52%	20	87%	12	52%
TCS	16	70%	13	57%	16	70%
TCSint	23	100%	17	74%	23	100%

other hand, this table allows us to evaluate in relative terms the contribution of each viewpoint.

7.6 DISCUSSION

The aim of this work was to design a conceptual framework, following the exploration and integration of domain knowledge schemes, by focusing on the systematization of ontological knowledge. More precisely, the domain knowledge was related to water management from the environmental and administrative viewpoints. This work has addressed the integration of different types of knowledge facing the challenges associated with the management of environmental issues involving water resources. This can be positive for integrated environmental modeling in decision support systems.

The proposed AI-based method for knowledge evaluation and integration allows several comparisons between the schemas to be integrated (viewpoints) and the overall schema to assess the contribution of each schema. As the results show, it can be confirmed that the integrated ontology is the most supported in terms of evaluation parameters, both relational and conceptual.

7.7 CONCLUSIONS

The evaluation of existing knowledge is one of the premises to be taken into account in any project to integrate new knowledge. In the case study described in this work, the integration of knowledge from local stakeholders and scientists is promoted. Thus, the knowledge has been defined based on the different points of view of the experts involved, namely, politicians, biologists, and farmers. However, to check for the validity and reliability of different knowledge, an evaluation is required to verify whether different forms of knowledge are being incorporated. Or it can be tested whether actors representing the interests of the administrations responsible for managing the environmental problems referred to in this work have been selected. For this reason, it was emphasized that the different points of view gathered in this research should be interpreted, analyzed, and agreed upon by local community members and scientists themselves, rather than by external experts.

In this study, it has been shown that artificial intelligence has been useful for the automatic evaluation and integration of knowledge from various knowledge sources.

It is considered that knowledge integration processes can be improved in further research by promoting certain actions, such as designing more effective methods for classifying different epistemological beliefs to have input ontologies that provide valid knowledge. In addition, applying knowledge-sharing methods between local stakeholders and scientists and expanding the use of ontology learning methods will be explored in future research.

ACKNOWLEDGMENTS

We thank the University of Murcia for its support under the grant "Information System for Information Integration on Sustainable Water Management" within the Agroalnext project. This study formed part of the AGROALNEXT programme and was supported by MCIN with funding from European Union NextGenerationEU (PRTR-C17.I1) and by Fundación Séneca with funding from Comunidad Autónoma Región de Murcia (CARM).

REFERENCES

1. Davenport, T. H., & Prusak, L. 1998. *Working Knowledge*; Cambridge, MA: Harvard University Press.
2. Tsoukas, H., & Vladimirou, E. 2001. What is organizational knowledge? *Journal of Management Studies*, 38, 973–993. https://doi.org/10.1111/1467-6486.00268.
3. Metolo, A. F. 2024. *AI and Environmental Conservation: Where We've Come from and Where we're Going*. https://es.weforum.org/agenda/2024/03/ia-en-la-conservacion-ambiental-de-donde-venimos-y-hacia-donde-vamos/.
4. Walcazak, S. 2005. Organizational knowledge management structure. *The Learning Organization*, 12(4), 330–339. http://doi.org/10.1108/09696470510599118.
5. Aportela, I. M., & Ponjuán, G. 2008. The second generation of knowledge: A new approach to knowledge management. *Science of Scientific and Technological Information*, 39(1), 19–30.
6. KMCI. 2001. *A Brief Introduction to KMCI's Conceptual Frameworks as Taught in the CKIM and K-STREAM™ Programs*. www.kmci.org/media/Intro_to_KMCIs_Frameworks.pdf.
7. McElroy, M. W. 2003. *The New Knowledge Management: Complexity, Learning, and Sustainable Innovation*; Boston, MA: KMCI Press, Butterworth-Heinemann.
8. Carlile, P. R. 2004. Transferring, translating, and transforming: An integrative framework for managing knowledge across boundaries. *Organization Science*, 15, 555–568. https://doi.org/10.1287/orsc.1040.0094.
9. Nicolini, D., Mengis, J., & Swan, J. 2012. Understanding the role of objects in cross disciplinary collaboration. *Organization Science*, 23, 612–629. https://doi.org/10.1287/orsc.1110.0664.
10. Carlile, P. R., & Rebentisch, E. S. 2003. Into the black box: The knowledge transformation cycle. *Management Science*, 49, 1180–1195. https://doi.org/10.1287/mnsc.49.9.1180.16564.
11. Kogut, B. 2000. The network as knowledge: Generative rules and the emergence of structure. *Strategic Management Journal*, 21, 405–425. https://doi.org/10.1002/(SICI)1097-0266(200003)21:3<405::AID-SMJ103>3.0.CO;2-5.
12. Nerkar, A., & Roberts, P. W. 2004 Technological and product-market experience and the success of new product introductions in the pharmaceutical industry. *Strategic Management Journal*, 25, 779–799. https://doi.org/10.1002/smj.417.

13. McDowell, W. C., Peake, W. O., Coder, L., & Harris, M. L. 2018. Building small firm performance through intellectual capital development: Exploring innovation as the "black box". *Journal of Business Research*, 88, 321–327. https://doi.org/10.1016/j.jbusres.2018.01.025.

14. Acharya, C., Ojha, D., Gokhale, R., & Patel, P. C. 2022. Managing information for innovation using knowledge integration capability: The role of boundary spanning objects. *International Journal of Information Management*, 62, 102438. https://doi.org/10.1016/j.ijinfomgt.2021.102438.

15. Carlile, P. R. 2002. A pragmatic view of knowledge and boundaries: Boundary objects in new product development. *Organization Science*, 13, 442–455. https://doi.org/10.1016/j.ijinfomgt.2021.102438.

16. Nicolini, D., Mengis, J., & Swan, J. 2012. Understanding the role of objects in cross-disciplinary collaboration. *Organization Science*, 23, 612–629.

17. de Hoop, E. 2020. More democratic sustainability governance through participatory knowledge production? A framework and systematic analysis. *Sustainability*, 12(15), 6160. https://doi.org/10.3390/su12156160.

18. Sauermann, H. et al. 2020. Citizen science and sustainability transitions. *Research Policy*, 49, 103978. http://doi.org/10.1016/j.respol.2020.103978.

19. Brandt, P. et al. 2013. A review of transdisciplinary research in sustainability science. *Ecological Economics*, 92, 1–15. http://doi.org/10.1016/j.ecolecon.2012.04.017.

20. Loeber, A. et al. 2007. The practical value of theory: Conceptualising learning in the pursuit of a sustainable development. In *Social Learning Towards a Sustainable World: Principles, Perspectives, and Praxis*; Wals, A. E. J., Ed.; Wageningen, The Netherlands: Wageningen Academic Publishers, pp. 83–98.

21. Woodhill, J. 2010. Sustainability, social learning and the democratic imperative: Lessons from the Australian landcare movement. In *Social Learning Systems and Communities of Practice*; Blackmore, C., Ed.; London, UK: Springer, pp. 57–72.

22. Irwin, A. 1995. *Citizen Science: A Study of People, Expertise and Sustainable Development*; London, UK: Routledge.

23. Wynne, B. 1996. Misunderstood misunderstandings: Social identities and public uptake of science. In *Misunderstanding Science? The Public Reconstruction of Science and Technology*; Irwin, A., Wynne, B., Eds.; Cambridge, UK: Cambridge University Press.

24. Stilgoe, J., Lock, S. J., & Wilsdon, J. 2014. Why should we promote public engagement with science? *Public Understandong of Science*, 23, 4–15. https://doi.org/10.1177/0963662513518154.

25. Irwin, A. 2014. From deficit to democracy (revisited). *Public Understandong of Science*, 23, 71–76. https://doi.org/10.1177/0963662513510646.

26. Brown, P. 1992. Popular epidemiology and toxic waste contamination: Lay and professional ways of knowing. *Journal of Health and Social Behavior*, 33, 267–281. https://doi.org/10.2307/2137356.

27. Lövbrand, E. 2011. Co-producing European climate science and policy: A cautionary note on the making of useful knowledge. *Science and Public Policy*, 38, 225–236. https://doi.org/10.3152/030234211X12924093660516.

28. Hess, D. J. 2007. *Alternative Pathways in Science and Industry: Activism, Innovation, and the Environment in an Era of Globalization*; Cambridge, MA, USA: MIT Press.

29. Ottinger, G., Barandiarán, J., & Kimura, A. H. 2017. Environmental justice: Knowledge, technology, and expertise. In *The Handbook of Science and Technology Studies*; U. Felt, Fouché, R., Miller, C. A., Smith-Doerr, L., Eds.; Cambridge, MA, USA: The MIT Press.

30. Irwin, A. 2001. Constructing the scientific citizen: Science and democracy in the biosciences. *Public Understanding of Science*, 10, 1–18. https://doi.org/10.3109/a036852.

31. Corburn, J. 2005. *Street Science Community Knowledge and Environmental Health Justice*; Cambridge, MA, USA: The MIT Press.

32. Frickel, S. et al. 2010. Undone science: Charting social movement and civil society challenges to research agenda setting. *Science, Technology, & Human Values*, 35, 444–473. https://doi.org/10.1177/0162243909345836.
33. Raymond, C. M. et al. 2010. Integrating local and scientific knowledge for environmental management. *Journal of Environmental Management*, 91, 1766–1777. https://doi.org/10.1016/j.jenvman.2010.03.023.
34. Scoones, I. 1999. New ecology and the social sciences: What prospects for a fruitful engagement? *Annual Review of Anthropology*, 28, 479–507. http://doi.org/10.1146/annurey.antro.28.1.479.
35. Nowotny, H., Scott, P., & Gibbons, M. 2001. *Re-thinking Science: Knowledge and the Public in an Age of Uncertainty*. Oxford: Blackwell Publisher.
36. Berkes, F. 2004. Rethinking community-based conservation. *Conservation Biology*, 18(3), 621–630. https://doi.org/10.1111/j.1523-1739.2004.00077.x.
37. Folke, C. et al. 2005. Adaptive governance of social–ecological systems. *Annual Review of Environment and Resources*, 30, 441–473. http//doi.org/10.1146/annurev.energy.30.050504.144511.
38. Agrawal, A. 1995. Dismantling the divide between indigenous and western knowledge. *Development and Change*, 26(3), 413–439. https://doi.org/10.1111/j.1467-7660.1995.tb00560.x.
39. Agrawal, A. 2002. *Indigenous Knowledge and the Politics of Classification*; Oxford, UK: UNESCO.
40. Sillitoe, P. 1998. The development of indigenous knowledge. *Current Anthropology*, 39(2), 223–252. https://doi.org/10.1086/204722.
41. Nygren, A. 1999. Local knowledge in the environment-development discourse. *Critique of Anthropology*, 19(3), 267–288. https://doi.org/10.1177/0308275X9901900304.
42. Bruckmeier, K., & Tovey, H. 2008. Knowledge in sustainable rural development: From forms of knowledge to knowledge processes. *Sociologia Ruralis*, 48(3), 313–329. https://doi.org/10.1111/j.1467-9523.2008.00466.x.
43. Bruckmeier, K., & Tovey, H. 2009. *Rural Sustainable Development in the Knowledge Society*; Derbyshire, UK: Ashgate.
44. Velásquez, T., Puentes, A. M., & Guzmán, J. A. 2011. Ontologias: una técnica de representación de conocimiento. *Revista Avances en Sistemas e Informática*, 8(2), 211–216.
45. Van Heijst, G. et al. 1997. Using explicit ontologies in KBS development. *International Journal of Human-Computer Studies*, 46, 183–292. https://doi.org/10.1006/ijhc.1996.0090.
46. Nguyen, N. P. et al. 2018. Cross-functional knowledge sharing, coordination and firm performance: The role of cross-functional competition. *Industrial Marketing Management*, 71, 123–134. https://doi.org/10.1016/j.indmarman.2017.12.014.
47. Laniak, G. et al. 2013. Integrated environmental modelling: A vision and roadmap for future. *Environmental Modelling & Software*, 39, 3–23. https://doi.org/10.1016/j.envsoft.2012.09.006.
48. Oprea, M. 2018. A knowledge modelling framework for intelligent environmental decision support systems and its application to some evironmental problems. *Environmental Modelling & Software*, 110, 72–94. https://doi.org/10.1016/j.envsoft.2018.09.001.
49. Ceccaroni, L. et al. 2004. OntoWEDSS: Augmenting environmental decision-support systems with ontologies. *Environmental Modelling & Software*, 19(9), 785–797. https://doi.org/10.1016/j.envsoft.2003.03.006.
50. Chen, S. H. et al. 2012. Good practice in Bayesian network modelling. *Environmental Modelling & Software*, 37, 134–145. https://doi.org/10.1016/j.envsoft.2012.03.012.
51. Heller, U., & Struss, P. 1996. Transformation of qualitative dynamic models-application in hydroecology. In *Proceedings of the 10th International Workshop on Qualitative Reasoning*; AAAI Press, pp. 83–92

52. Cortés, U., Sánchez, M., & Ceccaroni, L. 2000. Artificial intelligenca and environmental decision support systems. *Applied Intelligence*, 13, 77–91. https://doi.org/10.1023/A:1008331413864.

53. Chau, K. W. 2007. An ontology-based knowledge management system for flow and water quality modelling. *Advances in Engineering Software*, 38, 172–181. https://doi.org/10.1016/j.advengsoft.2006.07.003.

54. Xiaomin, Z. et al. 2016. An onotology-based Knowledge modelling approach for river water quality monitoring and assessment. *Procedia Computer Science*, 96, 335–344. https://doi.org/10.1016/j.procs.2016.08.146.

55. Guarino, N. 1995. Formal ontology, conceptual analysis and knowledge representation. *International Journal of Human and Computer Studies*, 43(5/6), 625–640. https://doi.org/10.1006/ijhc.1995.1066.

56. Fridman-Noy, N., & Hafner, C. D. 2010. Ontological foundations for experimental science knowledge bases. *Applied Artificial Intelligence*, 14(6), 565–618. https://doi.org/10.1080/08839510050076972.

57. Pinto, H. S., & Martins, J. P. 2001. Ontology integration: How to perform the process. In *Proceedings of International Joint Conference on Artificial Intelligence*; Washington, USA.

58. Fernández-Breis, J. T., & Martínez-Béjar, R. 2000. A cooperative tool for facilitating knowledge management. *Expert Systems with Applications*, 18(4), 315–330. https://doi.org/10.1016/S0957-4174(00)00013-0.

59. Shaw, M. L. G., & Gaines, B. R. 1989. A methodology for recognising conflict, correspondence, consensus and contrast in a knowledge acquisition system. *Knowledge Acquisition*, 1(4), 341–363.

60. Euzenat, J. 1996. Corporate memory through cooperative creation of knowledge based systems and hyper-documents. In *Proceedings of the 10th Workshop on Knowledge Acquisition, Modelling and Management*; Banff, Canada, pp. 1–36.

61. Farquhar, A., Fikes, R., & Rice, J. 1997. The Ontolingua server: A tool for collaborative ontology construction. *International Journal of Human-Computer Studies*, 46, 707–727. https://doi.org/10.1006/ijhc.1996.0121.

62. Tennison, J., & Shadbolt, N. 1998. APECKS: A tool to support living ontologies. In *Proceedings of the 11th Workshop on Knowledge Acquisition, Modelling and Management*; Banff, Canada.

63. Fridman-Noy, N., & Musen, M. A. 1999. An algorithm for merging and aligning ontologies. In *Automation and Tool Support. 16th National Conference on Artificial Intelligence, Workshop on Ontology Management*; Orlando, FL.

64. Stumme, G., & Maedche, A. 2001. FCA-merge: A bottom-up approach for merging ontologies. In *JCAI '01—Proceedings of the 17th International Joint Conference on Artificial Intelligence*; Seattle, USA.

65. McGuiness, D. L., Fikes, R., Rice, J., & Wilder, S. 2000. An environment for merging and testing large ontologies. In *KR2000: Principles of Knowledge Representation and Reasoning*; Cohn, A., Giunchiglia, F., Selman, B., Eds.; San Francisco, USA: Morgan Kaufmann, pp. 483–493.

66. Fenández-Breis, J. T. 2023. *An Ontology Integration Environment for the Development of Knowledge Management Systems*; Murcia: UMU.

67. García-González, M. S. et al. 2023. An artificial intelligence-based model for knowledge evaluation and integration in public organizations. *Applied Science*, 13, 11796. http://doi.org/10.3390/app132111796.

8 An In-Depth Exploration of Predictive Justice with AI

Mitisha Gaur

8.1 INTRODUCTION

Artificial intelligence ("AI") systems become more accessible with each passing year since an increasing amount of time and resources is devoted to the AI revolution. While it is observed that the AI revolution is primarily enabled in the private sectors, owing to the competitive nature of the private industries, deep pockets, and the race for constant evolution and profitability, the use of AI systems is also being increasingly observed within various governments and judicial bodies ("deployers") across jurisdictions, with their deployment of AI systems that are tasked with performing judicial and quasi-judicial functions ("predictive justice"). These deployers are hyper-focused on being first adopters, and therefore, it is observed that their use of AI systems is oriented solely around technical serviceability of an AI system, thus often ignoring the importance of the usability and reliability of the AI system from the perspective of both the users of the AI systems on behalf of the government and judicial bodies ("deployer users") as well as the persons on whom the computational results of the said AI system are applicable ("impact population"). The issues in the practical usability and reliability of an AI system generally emanate from AI system providers, which may either be the deployers or the developers of the AI system from whom the deployers license the AI systems for their own use. A plethora of issues has been identified when it comes to the use of AI systems to perform judicial and quasi-judicial functions, such as technical soundness of the AI system and the dataset on which the said AI system has been trained, the legal red tape and appropriate classification of the computations of AI systems (such as electronic evidence, expert testimony, etc.) on which the deployer users rely to pronounce judgments and orders upon the impact populations, and the role of adherence to procedural law which binds judicial and administrative bodies in the use of AI systems for performing judicial and quasi-judicial functions. This chapter investigates the impact of predictive justice applications on persons who may or may not have been historically discriminated against, with such historical discrimination encoded in the dataset used to train a predictive justice system, and further, these groups of people are not well-equipped to understand, adequately comprehend, object, or seek redressal in the event that they suffer damages, liability, or other material impacts to their life and liberty on account of being included in the part of the impact population ("vulnerable groups"). Vulnerability has been identified as a layered concept instead of an absolute concept. Layered vulnerability works in tandem with risk classification and requires constant observation of the deployer users as well as the impact population vis-à-vis the predictive justice systems [1]. The tussle between deployer users and the vulnerable groups that are a part of the impact population is a three-pronged issue: (1) Deployer users are seldom technically proficient to perform human oversight functions when it comes to AI systems. (b) The impact population's vulnerable groups are unaware of the fact that they are being subjected to the use of AI systems and therefore are unable to object to such practices. (c) The deployers of AI systems do not have robust internal governance properties and function as a vacuum sans risk mitigation and grievance redressal plans.

This chapter traces the various projects across jurisdictions, with special focus on the jurisdictions within the European Union ("EU") with deployers who are working with predictive justice systems. The chapter subsequently looks to trace some common issues noted across jurisdictions and works toward identifying the solutions proposed as a part of applicable soft laws and the regulatory matrix governing AI systems as laid out under the EU's draft Artificial Intelligence Act ("AIA"), as well as the legal actions taken against the deployer's use of predictive justice systems on legal grounds based in administrative laws and legal procedure codes.

8.2 PREDICTIVE JUSTICE THROUGH THE LENS OF AUTOMATED DECISION-MAKING

In 1950, Alan M. Turing published his seminal paper titled "Computing Machinery and Intelligence," where he posed the central question "Can machines think?" [2]. In his paper, Turing discussed the definitions of the terms *machines* and *think* and concluded that since the terms cannot be adequately defined, the focus must be drawn away from whether a machine can think but rather to if it can imitate the way a human functions. This led to the rise of the imitation game, which is what a machine must win in order to be proven to be a *thinking entity* and eventually named the *Turing test*, which has become the backbone of research in the field of artificially intelligent machines. A few years later, in 1956, John McCarthy coined the term *artificial intelligence* to define a system which amplifies the inherent intelligence and knowledge possessed by human beings. Over the last few decades, the use of AI has evolved from working as a basic algorithm performing simple functions, such as participating in a digital game as a player or proving a theorem, to its recent uses as a tool for human intelligence augmentation. A majority of the use cases of AI as a tool for intelligence augmentation has been enabled by automated decision-making technologies ("ADMT"), which are focused on simulating, emulating, or predicting results and decision-making capabilities of human beings based on large volumes of historical data. Automated decision-making ("ADM") is the process of taking a decision backed by insights garnered by automated means by relying on ADMTs, such as advanced analytics, machine learning, and deep learning, therefore ensuring minimal human involvement. These ADMs can be based on a variety of uniquely arranged datasets which may be composed of factual data, as well as on digitally created profiles or inferred data [3]. A few instances of ADM being applied to carry out crucial tasks with direct impact on the society include predictions regarding criminal recidivism, decisions pertaining to awarding loans to individuals, and conducting AI-powered aptitude tests as a part of a recruitment process. The use of ADM across certain scenarios, such as to carry out material decision-making and intelligence augmentation functions in law enforcement, judicial systems, administrative bodies, and employment scenarios ("high-risk scenarios"), has garnered both popularity and the attention of the regulators, which can be attributed to the material impact that reliance on these ADM systems has on the impact population. The unregulated and unsupervised use of ADM in these scenarios may lead to a direct adverse impact on the life, freedoms, and liberties of the impact population. Therefore, it is expected that these ADM systems shall be adequately designed and regulated to minimize a variety of different risks across various arenas in the society, namely: (1) individuals (such as, but not limited to, discriminatory practices based on gender, race, financial situations, unfair practices, loss of autonomy, etc.), (2) the economy (unfair commercial and regulatory practices, limited access to markets, etc.), and (3) society as a whole (manipulation, threat to democracy, etc.) [4].

The use of ADM in judicial and quasi-judicial ecosystems, such as courts, tribunals, administrative bodies, etc., better known as predictive justice, leads to a series of ethical, social, and legal enforceability-based questions, such as the fairness metrics which are required to be met by predictive justice applications, trustworthiness of predictive justice systems, transparency, and legal admissibility in judicial proceedings as per the applicable rules of evidence, which usually hinges on the transparency metrics of a predictive justice system [5]. As discussed previously, the use of ADM in high-risk scenarios is a tall order to meet and, therefore, is subjected to a combination of hard law and soft law requirements across jurisdictions, which are a combination of risk classification of the predictive justice systems, disclosure requirements, internal compliance policies, deployment policies of the predictive justice systems, as well as the grievance redressal mechanisms made available for the impact population.

8.3 PREDICTIVE JUSTICE APPLICATIONS ACROSS JURISDICTIONS

The first instances wherein judicial decisions were statistically predicted can be traced back to 2004, wherein a combined study titled "The Supreme Court Forecasting Project" ("SCFP"), conducted by students at the University of Pennsylvania, Washington University, and the University of California–Berkeley (United States), used a statistical model wherein the results of the model far surpassed the predictions made by skilled independent legal professionals. The SCFP focused on predicting the outcome of every case argued in 2002 in front of the United States Supreme Court. The statistical model employed by the SCFP was pitted against independent and experienced legal experts. The statistical model predicted 75% of the court's affirm/reverse results correctly, whereas the legal experts, in their collective capacity, were able to predict only 59.1% of the decisions correctly. However, we do observe a divide wherein the statistical model performed well in the cases pertaining to economic activity, while the legal experts performed comparatively better in cases pertaining to judicial powers. In the recent years, with the mandatory digitization of judicial and quasi-judicial decisions in an effort to accentuate access to justice and government intervention, we note that the collection of data for the training of predictive justice models has become quite straightforward. This leads to an increased ability of the deployers to develop and deploy predictive justice systems focused on carrying out intelligence augmentation functions in an effort to assist human beings or deployer users in performing various judicial and quasi-judicial tasks.

The Correctional Offender Management Profiling for Alternative Sanctions ("COMPAS") software, developed by Equivant (formerly Northpointe), is a criminal recidivism tool used in the courts of the state of Wisconsin, where the computational results of the COMPAS software were used to identify the risk of criminal recidivism based on a scoring mechanism which is based on fields such as criminal history, levels of education, stability in family life, etc. Equivant, in a 2009 study examining the predictive power of its COMPAS scoring system [6], defines *recidivism* as "a finger-printable arrest involving a charge and a filing for any uniform crime reporting (UCR) code."

COMPAS was widely investigated for the presence of bias against the Black population in the United States, and a detailed report by ProPublica [7] stated the various risks and harms to the life and liberty of the Black population, which is clearly a vulnerable group within the impact population, and the failure of the criminal justice machinery which was caused because of the use of such a criminal recidivism tool. Further, it is noted that the United States arrests and charges far more people than does any other country, a disproportionate number of them Black. Additionally, it has been noted and widely reported that for over 200 years, the responsibility to make key decisions in the criminal justice systems from the pretrial release stage to the sentencing stage to the parole stage has been tasked to human decision-makers, who are often guided by their instincts and personal biases [5]. Therefore, not only is there a presence of a historically biased dataset on which COMPAS has been trained, but there are also no human checkpoints for oversight at the time of data collection and feeding, which leads to the promulgation of a historical bias and further propagates the discriminatory patterns against the vulnerable groups within the impact population. Further, ProPublica, in its investigation report, noted that the data points required for the functioning of COMPAS, which are filled out in the form of a questionnaire, at the time of the arrest consist of detailed data, consisting of a total of 137 questions, which include information about family composition, childhood experiences (such as whether the parents of the accused person are divorced), social circle of the accused persons, as well as details about their personal and family life [8]. It has been noted that these 137 questions, or eventual COMPAS CORE data points, are often collected or filled out by secondary sources, such as arresting officers or acquaintances of the accused; therefore, they cannot be deemed to be a correct representation of facts. Further, some obvious errors, such as misquoting birth dates of the accused, were observed, which causes innocent correlation between the profiles of the accused and consequently had a direct and material impact on their eventual risk classification [7].

Further, following the indications of bias in the COMPAS system, the Wisconsin Supreme Court in the United States, in the case of *State vs. Loomis* [9], required a mandatory warning to be presented to the deployer as well as the deployer user before the use of algorithmic risk assessment in criminal sentencing, as many private companies are involved in the compilation of the presentencing investigation reports ("PIR"), and therefore, the risk of bias or failed fairness metrics is rampant in such risk-scoring practices.

In the Netherlands, the Systeem Risico Indicatie, or the SyRI algorithm, which was used by the taxation authorities to predict child welfare benefits fraud, targeted persons residing in low-income neighborhoods who were usually immigrants and slapped large fines on thousands of people. The SyRI scandal is another ripe example of predictive justice systems unfairly discriminating against vulnerable groups within the impact population. In its investigation report titled "Unprecedented Injustice" dated December 17, 2020, the Dutch Parliamentary Interrogation Committee working on the Parliamentary Questioning on Childcare Allowance categorically stated that the primary reasons for the deployment and reliance on the discriminatory Systeem Risico Indicatie, or the SyRI algorithm, have not solely been attributed to the failure of the algorithmic architecture of the system but as a deployer failure owing to the systemic top-down failure by the Dutch legislators and government officials to consider and protect the fundamental rights of non-discrimination and good governance of their citizens.

A similar incident was noted in the Swedish municipality of Trelleborg, which, owing to the Swedish government's agenda to embrace a digital transition of the public sector toward "digital first" and digital by default, caused them to automate various government functions, such as the Trelleborg municipality's effort to introduce fully automated decisions on applications for social benefits. The turning point in the processing of applications for social benefits was observed in 2017, when the number of persons who no longer rely on social benefits increased to 450 [10]. This was also the year that the predictive justice system was first introduced in the municipality of Trelleborg. This stark impact on the welfare provisions has been critically discussed across Sweden, wherein the very legal premise for the automation of a function which has a serious impact on the life and well-being of natural persons being delegated to an ADM system has been called into question [11]. Another point of concern against the use of the predictive justice system by the municipality of Trelleborg is the lack of transparency regarding the inner workings of the predictive justice algorithm, which further weakens the role of public administration institutions envisaged to take on the crucial role of an intermediary between the state and its citizens [10]. Thus, the use of predictive justice systems, which resulted in the denial of social benefits to persons who may be unemployed, owing to a plethora of reasons, such as health issues, physical and mental disabilities, inadequate levels of education, etc., is a decided targeting of vulnerable groups amongst an impact population in a scenario wherein the use of a predictive justice model may cause a direct impact on the social well-being, health, and quality of life of persons.

There has been active litigation regarding transparency requirements pertaining to the use of predictive justice by the municipality of Trelleborg. A journalist, Frederik Ramel, after several attempts to gain access to the source code of the predictive justice system being used by the municipality of Trelleborg, and also contacting the Danish company which is the developer of the predictive justice system of the Trelleborg municipality, filed an appeal before the Administrative Court of Appeal, arguing that the source code of the software used within the Trelleborg municipality's predictive justice system should be made publicly available as it falls under the Swedish principle of public access to official records. The Court allowed his appeal and upheld his request for access to the source code while citing the Swedish principle of public access to official records [12]. This judgment by the Swedish Administrative Court of Appeals has set a robust precedence when it comes to how predictive justice systems should be treated in terms of transparency requirements.

8.4   OBSERVED ISSUES WITH PREDICTIVE JUSTICE SYSTEMS

The delicate task of simulating human jurisprudence assigned to predictive justice algorithms is faced with a plethora of issues. The primary two issues this chapter discusses are (1) the eternal tussle between deliberative justice and predictive justice [13] and (2) the lack of transparency regarding the inner workings of the predictive justice system [14].

8.4.1   PREDICTIVE JUSTICE VS. DELIBERATIVE JUSTICE

Social trust is a crucial currency in the judicial and quasi-judicial systems, and the judiciary relies on public support to maintain its independence, especially where the independence of the judiciary is under attack in many countries [15]. Legal reasoning, attached to the judicial bodies, can be understood as a step in the *direct fitting of law*, through a cycle of variation, selection, and retention, to its social context [16].

Forecasting requires more than statistical correlation, as observed in predictive justice systems, if it works well [13]. There are many challenges that plague the use case of predictive justice, and these have been brought to the fore across the years. One of the main arguments which have been visited time and again is the reliance on predictive justice versus reliance on deliberative jurisprudence.

Many scholars, based on the study of American realism, cite the use of predictive justice as perfectly acceptable against the background of Justice Oliver Wendell Holmes's prediction theory of law, wherein he goes on to state the following:

> When we study law, we are not studying a mystery but a well-known profession. We are studying what we shall want in order to appear before judges or to advise people in such a way as to keep them out of court. The reason why it is a profession, why people will pay lawyers to argue for them or to advise them, is that in societies like ours the command of the public force is entrusted to the judges in certain cases, and the whole power of the state will be put forth, if necessary, to carry out their judgments and decrees. People want to know under what circumstances and how far they will run the risk of coming against what is so much stronger than themselves, and hence it becomes a business to find out when this danger is to be feared. The object of our study, then, is prediction, the prediction of the incidence of the public force through the instrumentality of the courts. A legal duty so called is nothing but a prediction that if a man does or omits certain things, he will be made to suffer in this or that way by judgment of the court—and so of a legal right. Law should be defined as a "prediction" of how most courts [17].

However convincing the argument of Justice Holmes may be, the advocates in favor of deliberative jurisprudence have many arguments stacked against the use of predictive jurisprudence, especially in the face of the usage of advanced tools such as AI. The fundamental argument used by the deliberative justice school of thought is focused on the jurisprudential aspects of predictive justice, which remains that the mandate of a court, whichever level it may be, is to deliberate on the set of facts presented before it against the backdrop of the law while providing a specific reasoned explanation for the judgment and the deliberations. The use of AI in predictive justice has many issues which attack the heart of the principles of natural justice.

The principles of natural justice require the adjudicating authority to provide a well-reasoned order; however, in most AI-based applications, there is a lack of adequate understanding of how the algorithm has arrived at a particular conclusion or, in the case of predictive justice, a judgment. This, apart from obliterating the requirement of a well-reasoned order, also dismantles the right to appeal, which has been granted to persons against whom an order has been issued, since the AI application cannot explain how it reached a specific conclusion, the grounds on which a judgment may be appealed are not clear [18]. There have been efforts to boost explainability in AI-based predictive justice systems, which can bring about the required transparency needed to operationalize these models; however, these are an exception and not the rule.

The second and another important principle of natural justice which may be eliminated by the use of predictive justice is the *rule against bias*. A sound predictive justice system requires a large dataset, which is the key element in deciding the sound functioning of the platform. A dataset that is comprised of biased data leads to algorithmic bias, as it provides a biased result [19].

Further, we have also observed cases where the reliance on the algorithmic computations of a predictive justice system may cause a habitual error in the judgments, which can only be understood as automation bias. One such example can be observed in the case of using risk assessment tools incorporated in the various aspects of the criminal justice system, such as in sentencing, evidence considerations, etc. [20].

8.4.2 TRANSPARENCY IN PREDICTIVE JUSTICE SYSTEMS

The pressure to streamline administrative, quasi-judicial, and judicial functions across governments has come to a head with the wide-scale algocratic endeavors, such as the adoption of AI systems, as has been observed in many jurisdictions [21, 22]. These predictive justice systems have been deployed for performing simple governmental functions which require little or no expertise, such as information collection, file management and retrieval, data grouping, etc. [23]. These tasks are considered low-risk tasks as they have little or no impact on the life, liberty, and well-being of the citizens that are subjected to the outcomes of these tasks and therefore may be fit to function with minimal or no human oversight [24]. However, there has been an increased adoption of predictive justice systems to carry out complex, high-risk tasks which use profiling based on a combination of personal as well as non-personal data, such as the use of AI for administrative investigations; the use of AI for carrying out the first-instance adjudication for small claims or minor offences, as well as minor civil wrongs, such as the imposition of fines for traffic violations, adjudication of rental disputes, adjudication of disputes pertaining to financial instruments, etc.; the use of AI in predictions pertaining to recidivism rates for convicted criminals; etc.

[25]. As noted in the preceding sections, high-risk predictive justice systems have been observed to be plagued with algorithmic discrimination that results in material impact on the life and liberty of vulnerable groups amid the impact population. This discriminatory impact on vulnerable groups has the potential to rupture the societal fabric, therefore forcing persons in vulnerable groups to the fringes of society. Therefore, the use of an AI black box model contributes directly toward algorithmic discrimination [26], especially since, in most cases, deployer users do not have the right level of training to use or rely upon predictive justice systems. The inherent opacity of the black box AI model and the inability to timely ascertain the presence of algorithmic discrimination cause irreparable damage against the citizens [27]. Further, these factors make the degree of discrimination faced by the subject tough to ascertain and therefore contributes toward the inability to make adequate reparations to the victims of such algorithmic discrimination [28].

In this context, the use of black box AI models to carry out these high-risk algocratic functions in the public administration ecosystems is a grave cause for concern since it has been observed that, in most cases, the computations of these black box models are relied upon blindly by the user to make decisions, and this inherent inability of the user to understand the details of the inner workings of the black box [26]. Further, in scenarios where an explanation or interpretation is not readily available, along with the computational result by a predictive justice system, there is an inherent chance of the deployer user (including deployer users trained in the use of AI systems) to succumb to automation bias and assume that the black box predictive justice system is performing a computation which is *de facto* fair, albeit not transparent [29]. Additionally, in scenarios where the deployer users are trained to question and investigate the computations of black box AI systems, the time and effort required to do that are counterproductive to the bottleneck elimination promise which the automation of government and administrative functions through the use of AI systems aims at offering.

8.5 SOLUTIONS

8.5.1 RESPONSE TO THE CHALLENGES IN PREDICTIVE JURISPRUDENCE: A SOFT LAW APPROACH

8.5.1.1 The European Union Agency for Fundamental Rights

The inadvertent permeation of AI-based predictive analysis tool in the legal system has been recognized by many countries across the world. The European Union ("EU") has been the harbinger of change with not just adopting the technology but also putting in place the necessary checks and balances required to effectively amalgamate the use of predictive jurisprudence with society. The EU Agency for Fundamental Rights ("FRA") published a report [30] in 2020 under the directorship of Michael O'Flaherty titled "Getting the future right: AI and fundamental rights" ("FRA Report").

The FRA Report discusses at length the various functions which AI performs that were previously only possible through natural persons, such as adjudication, analyses, marketing, etc., and while it recognizes this new wave of changes in the society in the face of AI-led development, it poses an acute awareness for the upholding of the fundamental rights of natural persons. The report is an amalgamation of a study by the FRA wherein over 100 public administration officials, private company staff, as well as a diverse pool of experts, which included a wide array of supervisory and oversight authorities, non-governmental organizations, and lawyers working in the field of AI, were interviewed. The FRA Report presents notable examples of companies and public administrations in the EU using, or trying to use, AI, while also discussing the numerous potential implications for fundamental rights and investigating whether and how those using AI are considerate toward the fundamental rights of the AI's impact population.

The FRA Report specifically focuses on access to justice when discussing the impact of the current use of AI on selected fundamental rights. The Charter of Fundamental Rights of the European Union ("Charter"), under Article 47, lays down the *right to an effective remedy and to a fair trial*. Article 47 of the Charter is the most cited article during legal proceedings and reads as follows:

> Everyone whose rights and freedoms guaranteed by the law of the Union are violated has the right to an effective remedy before a tribunal in compliance with the conditions laid down in this Article.
>
> Everyone is entitled to a fair and public hearing within a reasonable time by an independent and impartial tribunal previously established by law. Everyone shall have the possibility of being advised, defended and represented.
>
> Legal aid shall be made available to those who lack sufficient resources in so far as such aid is necessary to ensure effective access to justice.

The sentiments of Article 47 of the Charter are core to the fundamental rights of natural persons across a majority of countries, with a similar rendition of Article 47 guiding the legal proceedings in most countries, upholding the impermeable nature of fundamental rights that a nation accords to its citizens, with the same being recurringly imposed by courts across various jurisdictions. Further, the Charter also includes specific fundamental rights which are core to the judicial process, such as the right to a fair trial (particularly the right to a natural judge established by law, the right to an independent and impartial tribunal, and equality of arms in judicial proceedings) and, where insufficient care has been taken to protect data communicated in open data, the right to respect for private and family life [31].

Continuing, the FRA Report states that *the right to an effective remedy also covers decisions taken with the support of AI technologies*. Additionally, the EU data protection law reconfirms that the right to an effective judicial remedy must be provided in relation to decisions by the controller or the processor [32], as well as the supervisory authority [33]. Therefore, we note that the data processed by AI-driven technologies are no exception.

The FRA Report also notes the opacity of AI applications as a cause of concern for reliance on AI-based predictive jurisprudence technology [34] and states the following:

> One prominent concern is the lack of transparency in the use and operation of new technologies. Algorithmic decision making is notoriously opaque: data collection,

algorithm training, selection of data for modelling or profiling, the situation around individual consent, effectiveness and error rates of the algorithm and other aspects are often not transparently reported.

8.5.1.2   The European Commission for the Efficiency of Justice

In its 31st plenary meeting, the European Commission for the Efficiency of Justice ("CEPEJ") adopted the *European Ethical Charter on the Use of Artificial Intelligence in Judicial Systems and Their Environment* [35] ("CEPEJ AI Charter").

The CEPEJ AI Charter is aimed at guiding private and public sector organizations that are working toward the development and deployment of AI applications which involve the processing of judicial data, which includes interim orders, appeal petitions, and judgments by courts and tribunals. The CEPEJ AI Charter makes a distinction between the judicial decision processing for civil matters and criminal matters and states the following:

> Judicial decision processing by artificial intelligence, according to their developers, is likely, in civil, commercial, and administrative matters, to help improve the predictability of the application of the law and consistency of court decisions, subject to compliance with the principles set out below. In criminal matters, their use must be considered with the greatest reservations in order to prevent discrimination based on sensitive data, in conformity with the guarantees of a fair trial.

The CEPEJ AI Charter lays down five fundamental principles [35] to guide the development, as well as the deployment of AI-backed technology that processes judicial decisions and related data. These provide a much-required benchmark for organizations, whether in the private or the public sectors, which are involved in the sphere of predictive jurisprudence ("predictive jurisprudence principles"). These are as follows:

1. *Principle of respect for fundamental rights.* This highlights the responsibility to ensure that the design and implementation of artificial intelligence tools and services are compatible with fundamental rights.
2. *Principle of non-discrimination.* This is aimed at specifically preventing the development or intensification of any discrimination between individuals or groups of individuals.
3. *Principle of quality and security.* This principle works with regard to the processing of judicial decisions and data, using certified sources and intangible data with models conceived in a multi-disciplinary manner, in a secure technological environment.
4. *Principle of transparency, impartiality, and fairness.* This principle is aimed at making data processing methods accessible and understandable, to authorize external audits.
5. *Principle of "under user control."* This principle is designed to ensure the preclusion of a prescriptive approach and to ensure that users are informed actors and in control of their choices.

8.5.1.3   Need for Adoption of the Core Principles of Natural Justice in Predictive Justice

The legal procedural requirements across jurisdictions—whether civil law, common law, etc.—commonly abide by the principles of natural justice, which are (1) the rule against the presence of bias, (2) the rule of pronouncement of a reasoned order by the adjudication authority, (3) the rule against inordinate delay in adjudication, (4) the rule for the ability of a person to make legal representation in front of the adjudication authority, and (5) the rule for adequate notice to be provided to a person to prepare for the legal proceedings initiated against them [36].

This brings us to an important component in all predictive justice–based AI applications: a degree of explainability. Explainable AI ("XAI") has made many developments in recent times, and a degree of explainability in a predictive justice application is crucial inasmuch as it allows for natural persons to readily rely on them since they understand the reasoning behind the computational results of the AI [37]. The use of XAI as a core design tenet (which contributes to the accountability of the system) also enables the predictive justice application to function with a higher degree of reliability by the deployer use and directly affects trust vis-à-vis the predictive justice system and the impact population.

The inculcation of the PNJ focused on due process of law, that is, the rule for the ability of a person to make legal representation in front of the adjudication authority, is proposed as another core tenet to be inculcated by the deployer while enforcing a predictive justice system.

8.5.1.4   Predictive Jurisprudence and the Rule of Pronouncement of a Reasoned Order by the Adjudication Authority

A long-standing and well-founded concern while employing AI in the case of predictive justice is that most AI systems operate as black boxes, thereby impeding the possibility to retrace the reasoning behind a particular computational result that they provide. This inability to conclusively trust and rely upon a predictive justice system owing to its opacity is a major roadblock which stands in the way of augmentation of judicial functions. The ability to provide a reasoned order, albeit mechanically generated based on statistical insights, and the ability of a deployer user to rely on the input of a predictive justice system are based on the ability of the internal computations of the machine to be explained. This is where the explainability of AI takes center stage in the current scenario vis-à-vis vulnerable groups in impact populations.

Explainability (or "interpretability" [38]) in the context of social sciences can be understood as a concept centered on the expectation that a machine learning model (which is usually the type of model in use when we discuss AI applications) and its output can be explained in a way that is adequately comprehensive to a human being. In keeping with this definition, *interpretability* in the context of legal sciences is defined as the "ability to explain or to present in understandable terms to a human" [39].

Certain categories of algorithms, which include the more traditional machine learning algorithms, tend to be more readily explainable; however, they have been observed to be less efficient. Explainability is a core component in ensuring the creation of trust and reliability in AI, especially in the case of human-centered AI ("HCI"), such as predictive jurisprudence. However challenging the use of artificially intelligent predictive jurisprudence tools may be, an emerging solution is the use of XAI, which is achieved through the modeling of the AI architecture in a way which includes the processing of components that make it easier for the user to deduce the basis of the computation and rely on the results.

The reliance of predictive jurisprudence on XAI holds great promise as it looks to allow deployer users to comfortably rely on the computation of an XAI-based tool. This can also allow persons working in the predictive jurisprudence sphere to create an AI architecture which utilizes specific parameters to train the AI [26].

XAI also enables the AI to not only be a reliable tool but also be sufficiently robust and transparent, to be relied upon to carry out automated tasks which would earlier require constant human oversight [19]. Further, the creation and adoption of the XAI is currently left to private players; however, most HCIs, such as predictive jurisprudence and other applications, are usually deployed in the public domain. Therefore, it is key to ensure the involvement of public authorities in the development and deployment of XAI as an HCI tool for legal, judicial, and administrative functions. The use of opaque algorithms can destabilize the senses of transparency and fairness, which are focal to the functions of judicial authorities and public offices [40].

8.6   VULNERABLE GROUPS, PREDICTIVE JUSTICE, AND THE EU's ARTIFICIAL INTELLIGENCE ACT

Owing to its classification as a high-risk AI system under the provisions of the AIA, the providers of predictive justice applications are mandated to comply with certain regulations and good practices. The providers, which include the developers, that is, the companies who create predictive justice applications, as well as the deployers, that is, the persons (in this case, judicial and government departments) who deploy predictive justice applications across their departments and are vicariously responsible for the actions of the deployer users, are required to create robust internal governance policies and liability structures which apply to predictive justice applications [41]. These policies include grievance redressal mechanisms and requirements to conduct frequent and thorough technical assessments of the predictive justice systems. Further, the developers of predictive justice systems are required to ensure that these systems are placed on the market, that is, licensed to the deployers, only if they comply with the mandatory requirements imposed on them through the provisions of the AIA. The recognition of AI providers (developers and the deployers) as central stakeholders under the AI Act also allows for the creation of a deeply collaborative ecosystem within which providers are expected to continue to collaborate vis-à-vis the refining, monitoring, and maintenance efforts across the life cycle of the predictive justice system.

Since predictive justice systems are deployed by state actors, such as judicial authorities, administrative bodies, and law enforcement, therefore, under the scheme of the AIA, these state actors are obliged to carve out technical and organizational measures in order to ensure the smooth and reliable functioning of the predictive justice systems. These measures may range from periodic technical audits to third-party audits of the AI systems; compliance with cybersecurity standards; the creation of internal governance policies for the AI systems, which include a dedicated internal emergency response mechanism, a user- and impact population–oriented grievance redressal mechanism, a user training guide, etc.; and any other deployer-centric obligations which may be mandated by law [42].

Under the AIA, the deployers of high-risk AI systems are required to carry out a fundamental rights impact assessment ("FRIA"). The FRIA, especially in the cases of deployers who are rolling out predictive justice applications, is a critical piece of the ethics-by-design and human-rights-by-design puzzle [43]. The FRIA is required to be context-specific and is expected to identify and assess the AI system's impact in the specific context of use. The aspects of AI system which are expected include a clear outline of the intended purpose for which an AI system is to be used, a clear outline of the intended geographic and temporal scope of the AI system [43], categories of natural persons and groups which are likely to be affected by the use of the AI system, verification that the said AI system is in compliance with the applicable laws, specific risks of harms which are likely to impact marginalized persons or vulnerable groups, reasonable foreseeable impact on fundamental rights of putting the high-risk AI system into use, a detailed plan as to how the harms and the negative impact on fundamental rights identified will be mitigated, etc.

The FRIA is an especially impactful tool to regulate predictive justice systems and protect vulnerable groups against unforeseen damages and injuries, as they are required in order to specify the risks to vulnerable groups and marginalized persons. Further, predictive justice systems are also required to adhere to the provisions pertaining to transparency requirements and human oversight, as have been detailed in the AIA [44].

The provisions under the AIA, coupled with the soft law approach comprising of disclosure requirements and governance principles which the EU takes under the FRA Report, as well as the CEPEJ Charter, create a robust governance matrix for the deployment of predictive justice systems by deployers and take us one step closer to a sustained reality of creating predictive justice systems which keep in focus the interests of vulnerable groups within the impact population.

8.7 CONCLUSION

Systems that are designed keeping in mind the challenges faced by vulnerable groups are designed to be safer for all. There have been countless incidents across jurisdictions all over the world wherein a simple automation effort has gone haywire, resulting in a direct adverse impact upon the life, liberty, and well-being of persons in the impact population. Of the persons in the impact population, the persons who are a part of the vulnerable groups, irrespective of where they fall on the spectrum of vulnerability, are the hardest-hit. It is observed that most vulnerable groups which are affected by algorithmic discrimination are prey to biased datasets, which lead to lack of fairness in predictive justice systems and the promulgation of historical bias, the lack of transparency in the AI system, inadequately qualified deployer users, the lack of internal policies governing the deployment of predictive justice systems by deployers, etc. There are various soft and hard law requirements, forged in principles of fair treatment of persons and good governance, which are available in the EU's legal landscape which encompass disclosure requirements on behalf of deployers, the ability to gain redressal against the harms suffered at the hands of deployers, the right to an explanation, and the enactment of grievance redressal mechanisms as well as core principles, such as the principle of respect for fundamental rights; the principle of non-discrimination; the principle of quality and security; the principle of transparency, impartiality, and fairness; and the principle of "under user control." The need to design a system which takes into account the hurdles faced by vulnerable groups within the impact population is crucial to ensure the adoption of predictive justice systems without fear of algorithmic discrimination or lack of transparency while, at the same time, ensuring that public faith in government and judicial machinery is maintained.

ACKNOWLEDGMENTS

This research was supported by the European Commission under the Horizon 2020 Legality Attentive Data Scientist Project (LeADS), grant number 956562.

REFERENCES

1. Malgieri, G., & Niklas, J. (2020). Vulnerable data subjects. *Computer Law & Security Review*, 37, 105415.
2. Turing, A. M. (2009). *Computing machinery and intelligence* (pp. 23–65). Springer Netherlands.
3. Regulation (EU) 2016/679 (General Data Protection Regulation).
4. Mingtsung, C., & Shuling, L. (2020, June). Research on the application of artificial intelligence technology in the field of justice. In *Journal of physics: Conference series* (Vol. 1570, No. 1, p. 012047). IOP Publishing.
5. European Parliament, The Committee on Civil Liberties, Justice and Home Affairs. (2021). *Artificial intelligence in criminal law and its use by the police and judicial authorities in criminal matters*. https://oeil.secure.europarl.europa.eu/oeil/popups/fiche-procedure.do?lang=en&reference=2020/2016(INI).
6. Ciccolini, J. (2018). *Actuarial injustice: Discrimination in crime prediction software* (Doctoral dissertation, Columbia University).
7. Angwin, J., Larson, J., Mattu, S., & Kirchner, L. (2016, May). *Machine bias*. ProPublica. www.propublica.org/article/machine-bias-risk-assessments-in-criminal-sentencing.
8. *Equivant (previously Northpointe), Sample COMPAS risk assessment, COMPAS CORE*. Made Available by Julie Angwin (ProPublica). www.documentcloud.org/documents/2702103-Sample-Risk-Assessment-COMPAS-CORE.html.
9. State Vs. Loomis, 881 N.W.2d 749 (Wis. 2016).
10. Kaun, A. (2022). Suing the algorithm: The mundanization of automated decision-making in public services through litigation. *Information, Communication & Society*, 25(14), 2046–2062.
11. Larasati, Z. W., Yuda, T. K., & Syafa'at, A. R. (2023). Digital welfare state and problem arising: An exploration and future research agenda. *International Journal of Sociology and Social Policy*, 43(5/6), 537–549.
12. Kaun, A. (2022). Suing the algorithm: The mundanization of automated decision-making in public services through litigation. *Information, Communication & Society*, 25(14), 2046–2062.

13. Bex, F., & Prakken, H. (2021). Can predictive justice improve the predictability and consistency of judicial decision-making? In *Legal knowledge and information systems* (pp. 207–214). IOS Press.
14. Rudin, C., & Radin, J. (2019). Why are we using black box models in AI when we don't need to? A lesson from an explainable AI competition. *Harvard Data Science Review*, 1(2), 10–1162.
15. Chatziathanasiou, K. (2022). Beware the lure of narratives: "hungry judges" should not motivate the use of "artificial intelligence" in law. *German Law Journal*, 23(4), 452–464.
16. Deakin, S., & Markou, C. (2020). Evolutionary interpretation: Law and machine learning. *Journal of Cross-Disciplinary Research in Computational Law*. University of Cambridge Faculty of Law Research Paper (Forthcoming).
17. Holmes, O. W. (1897). The path of the law. *Harvard Law Review*, 10.
18. Deeks, A. (2019). The judicial demand for explainable artificial intelligence. *Columbia Law Review*, 119(7), 1829–1850.
19. Lee, N., Resnick, P., & Barton, G. (2019). *Algorithmic bias detection and mitigation: Best practices and policies to reduce consumer harms*. Brookings Institution. United States of America. https://coilink.org/20.500.12592/k29pdg. 22 Aug 2024. COI: 20.500.12592/k29pdg.
20. Kehl, D., Guo, P., & Kessler, S. (2017). *Algorithms in the criminal justice system: Assessing the use of risk assessments in sentencing*. Responsive Communities Initiative, Berkman Klein Center for Internet & Society, Harvard Law School.
21. Lorenz, L. C. (2019). *The algocracy. Understanding and explaining how public organizations are shaped by algorithmic systems* (Master's thesis).
22. Fricano, A. (2020). "Algocracy": the decline of representative democracy. In *CEUR workshop proceedings* (pp. 93–101).
23. Mehr, H., Ash, H., & Fellow, D. (2017, August). *Artificial intelligence for citizen services and government* (no. 1–12). Ash Center—Democratic Government Innovation, Harvard Kennedy School.
24. Battina, D. S. (2017). Research on artificial intelligence for citizen services and government. *International Journal of Creative Research Thoughts (IJCRT)*, 2320-2882. ISSN.
25. Corrigan, C. C. (2022). Lessons learned from co-governance approaches–developing effective AI policy in Europe. In *The 2021 yearbook of the digital ethics lab* (pp. 25–46). Springer International Publishing.
26. Rudin, C. (2019). Stop explaining black box machine learning models for high stakes decisions and use interpretable models instead. *Nature Machine Intelligence*, 1(5), 206–215.
27. Sandvig, C., Hamilton, K., Karahalios, K., & Langbort, C. (2014). Auditing algorithms: Research methods for detecting discrimination on internet platforms. *Data and Discrimination: Converting Critical Concerns into Productive Inquiry*, 22(2014), 4349–4357.
28. Wachter, S., Mittelstadt, B., & Russell, C. (2021). Why fairness cannot be automated: Bridging the gap between EU non-discrimination law and AI. *Computer Law & Security Review*, 41, 105567.
29. Ahn, M. J., & Chen, Y. C. (2022). Digital transformation toward AI-augmented public administration: The perception of government employees and the willingness to use AI in government. *Government Information Quarterly*, 39(2), 101664.
30. The European Union Agency on Fundamental Rights. (2020). *Getting the future right: Artificial intelligence and fundamental rights*. https://fra.europa.eu/sites/default/files/fra_uploads/fra-2020-artificial-intelligence_en.pdf.

31. European Commission for The Efficiency of Justice. (2018). *European ethical charter on the use of artificial intelligence in judicial systems and their environment*, p. 16. www.europarl.europa.eu/cmsdata/196205/COUNCIL%20OF%20EUROPE%20-%20European%20Ethical%20Charter%20on%20the%20use%20of%20AI%20in%20judicial%20systems.pdf.

32. Law Enforcement Directive, Art. 54; and GDPR, Art. 79.

33. Law Enforcement Directive, Art. 53; and GDPR, Art. 78.

34. Ferguson, A. G. (2016). Policing predictive policing. *Washington University Law Review*, 94, 1109.

35. COUNCIL, O. E. (2018). *European ethical charter on the use of artificial intelligence in judicial systems and their environment*. European Commission for the Efficiency of Justice, Strasbourg.

36. Harlow, C. (2006). Global administrative law: The quest for principles and values. *European Journal of International Law*, 17(1), 187–214.

37. Gerards, J. H., Kulk, S., Berlee, A., Breemen, V. E., & Peters van Neijenhof, F. (2020). *Getting the future right: Artificial intelligence and fundamental rights*. Fundamental Rights Agency. https://doi.org/10.2811/774118.

38. Nori, H., Caruana, R., Bu, Z., Shen, J. H., & Kulkarni, J. (2021, July). Accuracy, interpretability, and differential privacy via explainable boosting. In *International conference on machine learning* (pp. 8227–8237). PMLR.

39. Wachter, S., Mittelstadt, B., & Russell, C. (2017). Counterfactual explanations without opening the black box: Automated decisions and the GDPR. *Harvard Journal of Law & Technology*, 31, 841.

40. Malgieri, G. (2019). Automated decision-making in the EU member states: The right to explanation and other "suitable safeguards" in the national legislations. *Computer Law & Security Review*, 35(5), 105327.

41. Schiff, D., Biddle, J., Borenstein, J., & Laas, K. (2020, February). What's next for AI ethics, policy, and governance? A global overview. In *Proceedings of the AAAI/ACM conference on AI, ethics, and society* (pp. 153–158). https://doi.org/10.1145/3375627.3375804.

42. Mökander, J., & Floridi, L. (2023). Operationalising AI governance through ethics-based auditing: An industry case study. *AI and Ethics*, 3(2), 451–468.

43. Janssen, H., Seng Ah Lee, M., & Singh, J. (2022). Practical fundamental rights impact assessments. *International Journal of Law and Information Technology*, 30(2), 200–232.

44. Calvi, A., & Kotzinos, D. (2023). Enhancing AI fairness through impact assessment in the European Union: A legal and computer science perspective. In *Proceedings of the 2023 ACM Conference on Fairness, Accountability, and Transparency (FAccT '23)* (pp. 1229–1245). Association for Computing Machinery. https://doi.org/10.1145/3593013.3594076.

9 Developments on Generative AI

Engy Yehia

9.1 DEVELOPMENT OF GENERATIVE AI

Artificial intelligence (AI) is a fast-growing domain that covers an extensive range of practical applications and continuing challenges in research. Intelligent software is relied upon to automate repetitive tasks, recognize spoken language or visual data, provide medical diagnoses, and assist in essential research [1]. In 1956, a group consisting of computer scientists put upward the proposition that computers could perform intelligently enough to participate in thinking and reasoning based on logic [2]. Systems that are based on rules and driven by experts, or data-driven machine learning training, can be used to achieve this [3]. The scientists' group described this principle as "artificial intelligence." In simple terms, AI is a discipline that concentrates on the automation of cognitive functions typically carried out by humans, while ML and DL are particular approaches to accomplishing this objective. This means that they fall under the domain of artificial intelligence, as shown in Figure 9.1, which depicts the relationships and development of AI domains. AI involves methodologies that do not rely on learning but instead concentrate on encoding explicit rules for all possible situations within a specific area of interest. These rules, which have been authored by humans, are derived from a priori knowledge pertaining to the specific subject and task that need to be completed [2]. For instance, if someone were to code an algorithm to provide investment decisions in the financial sector, programmers write a rules extracted from investment experts to produce investment suggestions, considering a number of variables, including market movements, historical data, and risk tolerance. In the last decade, artificial intelligence (AI) has attracted significant attention in both scientific and non-scientific areas. Machine learning (ML), deep learning (DL), and AI have been thoroughly discussed in several publications published, in both technology and non-technical publications [2].

9.1.1 KNOWLEDGE-BASED INFERENCE ENGINES

Various artificial intelligence initiatives have attempted to represent knowledge about the world using formal languages. Computers have the ability to automatically use logical inference rules to reason about assertions in formal languages. The approach referred to here is often known as the knowledge base approach in the field of artificial intelligence [1]. For example, knowledge-based inference engines are used to determine the most likely diagnosis in the medical sector; the engine uses the patient's symptoms

DOI: 10.1201/9781003501152-9

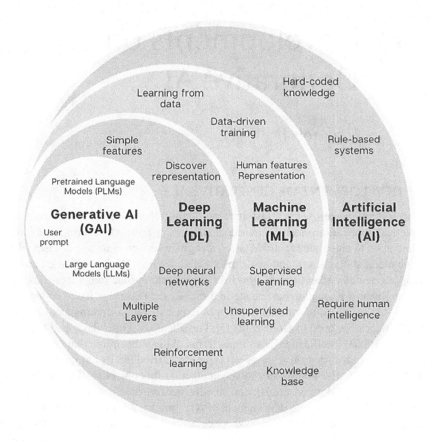

FIGURE 9.1   Development of artificial intelligence domains.

Source: Inspired from [1, 4, 5].

as input and applies a set of rules extracted from medical experts. The challenges observed by systems that depend on hard-programmed knowledge indicate that AI systems require the ability to gain knowledge by analyzing and identifying patterns in raw data. Machine learning refers to this particular power [1]. Complex problem-solving, commonly referred to as artificial intelligence (AI), relies on analytical models that produce predictions, rules, responses, suggestions, or similar results. Initial attempts to construct analytical models involved the use of explicit programming using manually generated rules (such as expert systems for medical diagnoses) [6].

9.1.2   Machine Learning

Machine learning, a specialized branch of artificial intelligence, is dedicated to developing algorithms that can independently perform tasks by learning from data, rather than relying on explicit programming [5]. Machine learning is a discipline that specifically concentrates on the process of AI by creating algorithms that accurately represent a given dataset. Unlike classical programming, where algorithms are implemented in a simple and straightforward way using well-known features, machine learning (ML) utilizes subsets of data to build algorithms that may utilize inventive or unique combinations of features and weights, which cannot be determined from first principles [2]. The development of machine learning has allowed computers to address complex problems that need the understanding of real-world information and to make decisions that may seem subjective. In the field of machine learning, there are three widely utilized learning methods, which are selected based on the specific problem and the data at hand: supervised learning, unsupervised learning, and reinforcement learning [7]. Commercial applications often use supervised learning to classify or forecast business data using algorithms trained on labeled datasets. The algorithm obtains the ability to make associations between input and output data and predict unexpected data. ML also uses unsupervised learning and reinforcement learning to acquire the optimal ability to make decisions, by engaging with the environment, and to continuously optimize cumulative rewards over a period of time [5]. Regression models, decision trees, instance-based algorithms, ANNs, and Bayesian approaches are among the ML algorithms available, depending on the learning task [7]. Machine learning (ML) may provide reliable and consistent findings by utilizing insight from prior calculations and identifying patterns from extensive databases. ML algorithms have shown effective in various areas, including fraud detection, credit scoring, natural language processing (NLP), and speech and image recognition [7]. Designing an appropriate set of features and providing them to a basic machine learning algorithm can successfully address various artificial intelligence problems [1]. For example, logistic regression, a basic machine learning technique, can find out if a patient has a disease or not. The functionality of these basic machine learning techniques heavily relies on the manner in which the data is represented [1]. For instance, when logistic regression is used to determine whether a patient has a disease or not, AI does not directly inspect the patient. Instead, the physician enters measurements like blood pressure and temperature. Many tasks make it hard to decide what features to extract. Various real-world artificial intelligence applications are challenging because various elements of variation affect all data we view. Most applications require us to separate and remove non-essential variation components. Extracting high-level, abstract features from raw data becomes a challenging task utilizing sophisticated, essentially human-level data interpretation. Deep learning deals with machine learning's main challenge by using simpler representations to express data representations [1].

9.1.3 Deep Learning

Deep learning algorithms are a distinct category of machine learning algorithms that focuses on discovering several layers of distributed representations. This approach is gaining popularity and has been extensively utilized in several classic artificial intelligence areas, including semantic parsing, natural language processing, transfer learning, computer vision, and more [8]. Deep learning enables the computer to produce more complicated concepts by incorporating simpler ones. An instance of a deep learning model is the feedforward deep network, sometimes referred to as a multilayer perceptron (MLP). A *multilayer perceptron* is a mathematical function that maps a given set of input values to their corresponding output values. The function is formed by integrating several elementary functions. Every time a unique mathematical function is used, it produces a new representation of the input [1]. Deep neural networks generally comprise multiple hidden layers arranged in complex hierarchical network designs. Deep neural networks have the ability to process raw input data and automatically discover a representation that is essential for the associated learning task. The networks' fundamental capability is widely referred to as deep learning [7]. According to this perspective on deep learning, it is not always necessary that all the information contained in a layer's activations represent the components that explain for the input.

To summarize, deep learning can be described as a form of machine learning that enables computer systems to enhance their performance through experience and data. Deep learning has demonstrated its potential in various software domains, such as natural language processing, computer vision, robotics, speech and audio processing, biology and chemistry, video games, search engines, finance, and online advertising [1].

9.1.4 GENERATIVE ARTIFICIAL INTELLIGENCE (GAI)

The field of machine learning has accomplished significant advancements over the years, leading to the emergence of a new branch called "generative AI." Machine learning mostly focuses on predetermined tasks, such as regression, classification, and prediction. These tasks involve training the model to assess and make judgments using a certain collection of input data [9]. Generative artificial intelligence (GAI) is a distinct field within AI that utilizes machine learning to generate novel and unique information by leveraging huge historical databases. Generative AI has a variety of possible applications, such as generating unique pictures, written content, and musical creations; speech recognition; natural language processing; and computer vision. Generative models can be used to create realistic images for video games, simulations, and virtual reality. This technique is achieved by employing a model that received training on a vast collection of instances and produces new instances that are close to the original dataset [10]. Figure 9.2 depicts the correlation across several AI fields and provides a broad overview of their operation. Generative AI employs advanced algorithms that evaluate data patterns to generate novel solutions. The progress in machine learning and deep learning, particularly in generative adversarial networks (GAN) and their modifications, has resulted in the creation of diverse systems across numerous areas. These include several models that convert text into different formats, such as ChatGPT for text-to-text, DALL-E 2 for text-to-images, models for text-to-music, models for text-to-video, and Alpha Code for text-to-code [9]. Generative AI exploits the capabilities of extensive language models (LLMs) by training them on vast text datasets and employing AI models with a substantial number of model parameters. ChatGPT, an AI model developed by OpenAI, is a widely recognized instance of a generative AI that utilizes language models (LLMs) [3]. *Foundation models* refer to deep learning models that received pretraining utilizing large datasets. These models can be used either as they are

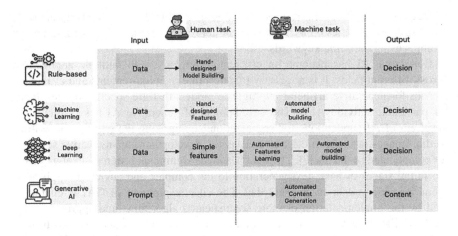

FIGURE 9.2 The difference between AI disciplines and a high-level schematic of how each works.

Source: Inspired from [1, 5, 7].

or with further refinement for a range of tasks, such as generating text and images or classifying them [11]. ChatGPT employs transformer neural networks that are pretrained on unlabeled extensive text corpora to gain an in-depth understanding of language patterns. At this phase, it is referred to as a substantial language model, also known as LLM. Subsequently, the model undergoes additional improvement to effectively respond to user prompts [3]. Generative AI has emerged as a significant technological breakthrough in the field of deep learning research in recent years [12]. Training a GAI model differs from training a traditional machine learning (ML) model since it involves semi-supervised learning. This approach combines distinct learning techniques by first using a small quantity of labeled data and then utilizing a large amount of unlabeled data (unsupervised learning). Current GAI models employ techniques like supervised fine-tuning (SFT), reward models, and reinforcement learning to guarantee that the model conforms to the goals and principles set by the creators [13]. This novel method enables the training of extremely extensive datasets necessary for GAI models without the necessity of intricate full tagging. The application system serves as a user interface for interacting with a GAI model. Prompting is an interactive approach and distinctive characteristic of GAI that allows users to engage with and provide instructions to GAI applications using natural language. This enables the generation of desired outputs, such as text, graphics, or other forms of information. Text-to-image programs employ textual instructions to define the intended visual aspects of a picture, while image-to-image applications rely on an input image to direct the generation process [5]. Generative AI models have been specifically constructed to provide probabilistic outputs, which means that the results they generate are not comparable in the same way as the deterministic outcomes of ML models. A GAI application will produce several outputs for the same input prompt, yet the outcomes are consistent and fulfil the request. Therefore, the process of creating a prompt that effectively achieves the desired result relies on a trial-and-error approach, such as by rephrasing textual prompts using the same keywords [14].

Generative AI has become an efficient technology with multiple applications across different fields. It is necessary to determine the requirements and evaluation metrics for generative artificial intelligence models that are specifically developed for particular activities [15]. The development of generative AI has resulted in a novel era of AI applications for organizations, comprising the creation of writing, media, images, art, and music, as well as the development of innovative solutions for complicated problems [9]. The field has made significant progress, evolving from the initial stages of basic text generation to the complex and advanced application scenarios observed in the present day. The current progress of GAI is characterized by the implementation of more sophisticated algorithms and approaches that facilitate the creation of novel and innovative content [16]. GAI has significantly transformed the world by facilitating the creation of novel experiences through the integration of virtual and physical realms. As the use of GAI expands in conjunction with the metaverse, it is being examined by scholars, researchers, and industrial communities to uncover its boundless potential. GAI is a prominent technology in both physical and virtual commercial platforms. It is utilized by several AI platforms, such as ChatGPT by OpenAI and Bard AI by Google [17].

9.2   CLASSIFICATION OF GENERATIVE AI MODELS

The emergence of GAI tools has attracted significant public attention, although their development has been ongoing for several years. The release of GPT-2 in 2019 represented the initial demonstration of GAI's potential to bring about extensive economic and societal changes. In 2023, there are many GAI tools that specialize in various tasks, such as generating text, images, videos, audio, and code [16]. Figure 9.3 provides a classification of the generative models that are frequently used in the field of generative artificial intelligence.

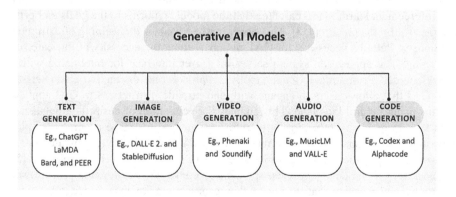

FIGURE 9.3   Classification of the generative AI models that are frequently used.

9.2.1 TEXT GENERATION MODELS

Text generation models focus on processing text inputs and producing text outputs to satisfy particular requirements [18]. For instance, the conversational agent ChatGPT uses the popular family of LLMs known as GPT (short for "generative pretrained transformer") to generate text [19]. GPT models can read and write content in several languages like a human being using natural language processing (NLP). They can also write creatively and effectively on nearly any subject, from a paragraph to an entire research essay. Even customer service chatbots, which use these models, can converse with clients in a manner approximating that of a human [20]. Large generative AI models that possess a comprehensive and diverse ability to simulate output in and across particular domains or data types are frequently referred to as foundation models [19]. OpenAI's well-known conversational agent ChatGPT uses the most recent generative AI model, GPT-4, which can produce text outputs from both image and text inputs [13]. Neural networks used to model and generate text data that often include three properties are referred to as (large) language models (LLMs). First, a large-scale sequential neural network is used by the language model. Second, auxiliary tasks are created to teach the neural network a representation of natural language without running the risk of overfitting, thereby pretraining the network through self-supervision. Third, massive text datasets (like Wikipedia, or even multinational datasets) are used in the pretraining. Language models have developed recently into billion-parameter-rich systems known as LLMs. BERT and GPT-3, which include 340 million and 175 billion parameters, respectively, are two examples of huge LLMs [19]. In the end, LLMs are simply stochastic parrots that repeat patterns from the data they have been trained on and guess which word matches in a phrase without making any ethical judgments. Consequently, a human with sufficient subject matter experience should always review the generated language because LLMs cannot be trusted to produce appropriate or factually accurate text [11]. ChatGPT, LaMDA, Bard, and PEER are examples of text generation models.

9.2.1.1 ChatGPT

Based on its training data, ChatGPT is a conversational assistant driven by generative AI that produces responses that closely mimic real human language [21]. OpenAI, the same company that created earlier iterations of GPT, is the creator of ChatGPT, which is powered by the underlying model GPT-3.5 [22]. With the ability to engage in conversational conversations with users and respond to their queries in natural language, ChatGPT is a potent model. The model is constructed using a transformer architecture, allowing it to efficiently handle large volumes of text and produce high-quality responses. ChatGPT uses a combination of supervised fine-tuning and reinforcement learning in its training process [18]. In January 2023, ChatGPT reached over 100 million users, making it the fastest-growing consumer application to date [17]. GPT models are typically trained using a two-stage process. Initially, they are trained by utilizing an extensive dataset of text sourced from the internet, with the objective of accurately predicting the next word. The models are subsequently fine-tuned with additional data, employing an algorithm known as reinforcement learning from human feedback (RLHF) to provide outputs that are preferred by human labelers [13].

TABLE 9.1

A High-Level Overview of Basic GPT Models for Text Generation

	Description	Release Year	Training Dataset	Parameters	Context Window	Model	Capabilities
GPT-1	GPT-1 changed the building of downstream task models through the adoption of the "pretrain and fine-tune" methodology.	June 2018	8 million web pages	117 million	512	Transformer-based pretrained language models (PLM)	GPT-1 model could perform a task without a prior example by generating the subsequent token by utilizing earlier tokens.
GPT-2	GPT-2 aims to forecast the subsequent word(s) in a given sentence.	February 2019	40 GB of web pages	1.5 billion	1,024	Transformer-based pretrained language models (PLM)	GPT-2 model could have further reduced perplexity and improve the understanding of human language by expanding the dataset used for initial training and increasing the length of the training period.
GPT-3	GPT-3 utilizes training-free in-context learning, which addresses downstream problems by leveraging the knowledge stored in the model parameters. GPT-3 has exhibited robust capabilities in "zero-shot" and "few-shot" learning across several tasks.	June 2020	45 TB of web pages	175 billion	2,048	Large language models (LLMs)	GPT-3 models have the ability to understand and create natural language. The GPT-3 model is far larger than the GPT-2 model, being 100 times its size. It has the ability to learn from extensive amounts of text data obtained from many sources, such as web pages, Wikipedia, and books.
GPT-3.5	GPT-3.5 models are created by refining GPT-3 models utilizing coding data and then adjusting them by either supervised fine-tuning (SFT) or reinforcement learning from human feedback (RLHF).	March 2022	Over 45 TB of web pages	Over 175 billion	16,000	Large language models (LLMs)	GPT-3.5 models have the ability to understand and generate both human language and computer code.
GPT-4	GPT-4 is a model that utilizes the transformer architecture and is trained in advance to forecast the subsequent token in a given document. The GPT-4 model offers the capability to handle both text and image input and provide textual outcomes.	March 2023	Extensive dataset comprising text and code	Over 1 trillion	128,000	Large language models (LLMs)	GPT-4 models can handle both text and image inputs. GPT-4 presents newer safety challenges compared to previous GPT models, which may produce biased and unreliable content.

ChatGPT can be used in some areas like chatbots and customer care systems, content generation, and language translation. Table 9.1 presents a brief description of the GPT family developed by OpenAI [4, 10, 13, 17, 23–25].

9.2.1.2   LaMDA

LaMDA, developed by Google, is a collection of conversational neural language models. In 2021, the first generation was revealed at the Google I/O keynote. In June 2022, LaMDA gained significant attention following the claims of Blake Lemoine, a Google employee, that the chatbot had developed sentience [10]. LaMDA is a customized neural language model designed for dialogue applications based on transformers. Unlike other language models, it was specifically trained on dialogues and has up to 137B parameters. The model received pretraining utilizing a vast amount of web content and public debate data, totaling 1.56 trillion words, and is among the largest pretrained language models. Because the model can handle complex conversations with a variety of answers, it is a helpful tool for many applications [18]. In February of 2023, Google launched Bard, a conversational AI chatbot driven by LaMDA [10].

9.2.1.3   Bard

Bard is a conversational GAI, similar to OpenAI's ChatGPT, that utilizes Google's LLM "LaMDA" technology. It has experienced extensive training on a large dataset, enabling it to generate text, do language translation, create diverse material, and provide relevant answers to inquiries. Bard has been utilized in diverse scientific applications, encompassing experiment design, data analysis, scientific writing, and literature evaluation [16].

9.2.1.4   PEER

PEER is a collaborative language model that has been trained on edit histories to cover the whole writing process. PEER, developed by Meta AI research, consists of four steps: plan, edit, explain, and repeat. PEER has the ability to compose preliminary versions, provide recommendations, suggest modifications, and offer justifications for its decisions [26]. The stages are continued until the text reaches a satisfactory state and no more revisions are required. The notion facilitates the partitioning of the process of writing a paper into multiple simpler subtasks. The approach is based on self-training, where models are used to complete missing data and then train other models using this artificially generated data. The primary source of training for the model is derived from the edit histories of Wikipedia. One problem of a retrieval approach is its failure sometimes to compensate for the frequent lack of citations and the loudness of remarks [18].

9.2.2   IMAGE GENERATION MODELS

Image generation models primarily rely on generative adversarial networks (GAN), which have the capability to generate realistic images or films [11]. Image creation apps create images in response to user input. By utilizing generative adversarial networks (GANs) or diffusion models called deep generative models (DGMs), artificial images are generated, which have practical applications in marketing, design, fashion, and other creative domains, serving as novel forms of visual art. Stable diffusion is an open-source text-to-image model that allows for the creation of images in various applications of GAI. In addition to the process of creating images from text, it is also possible to alter images using image-to-image systems. These systems may manipulate and expand images based on the user's instructions [5], the common image datasets employed for training models such as DALL-E or Stable Diffusion. Empirically, the models exhibit superior performance in relation to image quality [11]. Diffusion probability models formally represent the visual data by simulating the movement of data points via a hidden space, drawing reference from statistical physics. More precisely, they commonly employ Markov chains that have been trained using variational inference. Subsequently, they reverse the diffusion process to produce a realistic image. Commercial systems like DALL-E and Midjourney also employ diffusion probability models [19]. These algorithms generate diverse images based on the user's preferences, enabling them to select the topic, style, mood, setting, and other factors [22]. Artists and designers can utilize generative models to produce distinctive artworks, illustrations, and graphics. Content providers and bloggers can employ generative models to produce pertinent and visually appealing images to enhance their written content. Artists can explore unique visual representations by experimenting with different styles and customizing certain aspects. Teachers can utilize generative models to produce visually captivating teaching resources and materials [27].

9.2.2.1 DALL-E 2

OpenAI developed DALL-E 2 as an enhanced version of its predecessor, DALL-E [28]. DALL-E is an advanced artificial intelligence technology that employs natural language prompts to produce realistic images [21]. DALL-E 2 is capable of producing more realistic images at greater resolutions and possesses the capacity to seamlessly integrate ideas, features, and aesthetic. DALL-E 2 was trained using approximately 650 million image-text pairs obtained from the internet [28]. The CLIP (Contrastive Language-Image Pretraining) was trained using a variety of image and text combinations. CLIP utilizes natural language instructions to select the most relevant text extract based on an image and has recently emerged as an effective representation learner for images [18]. DALL-E 2 has the ability to produce novel synthetic visuals that align with given textual input (captions). The technology utilizes a deep learning model to generate images from text. DALL-E 2 can be utilized in various management domains, including product design, visual marketing, brand creation, and other applications [9].

9.2.2.2 Stable Diffusion

Stability AI established Stable Diffusion in 2022 with the main aim of producing aesthetically pleasing graphics based on textual descriptions. Furthermore, it can be employed for a multitude of other jobs, such as manipulating images and converting image formats. The model training process entails utilizing photos with dimensions of 512×512 pixels. This subset is a comprehensive aggregation of internet data collected by the German charity LAION [28]. The website API of Stable Diffusion can be employed to obtain access to it [18].

9.2.3 Video Generation Models

Video generation models involve the production of artificial videos, in particular, moving images with dynamic motion. Video clips can be created by either describing the content of the intended video footage using text or by applying the style and composition from a text or image prompt to an existing video. These opportunities enable the quick simple production and modification of videos using natural language and other methods. Video generation models are utilized in several areas, such as sales and marketing, where they are used for creating product marketing videos. They are also used in onboarding and education, where virtual avatars are employed in training movies [5]. Considering this, it is logical to generate films, which essentially consist of a sequence of images, based on text. In this section, we will discuss two models, Phenaki and Soundify, that have the ability to perform this task.

9.2.3.1 Phenaki

Phenaki is a video synthesis model created by Google Research capable of generating realistic videos based on textual prompts. The accessibility of this model's API is enhanced due to its availability on GitHub. Phenaki's unique quality lies in its capacity to produce films based on time-variable cues from open domains [18]. Phenaki consists of two primary components: an encoder-decoder model that compresses movies into discrete embeddings, and a transformer model that translates text embeddings into video tokens. Phenaki utilizes a pretrained language model called T5-XXL [29]. The encoder compresses videos, converting the initial tokens into embeddings. These embeddings are then passed through a spatial and temporal transformer. This technique generates diverse and chronologically coherent films using public domain prompts, even when the prompt involves a unique combination of ideas. The model has the capability to generate videos that are several minutes long, even though it is trained on videos that are only 1.4 sec in duration [18].

9.2.3.2 Soundify

Runway has created a method called Soundify that aids professionals in video editing by helping them locate and synchronize suitable sounds. The solution utilizes top-notch sound effect libraries and capitalizes on the zero-shot categorization capabilities of CLIP as a neural network. Soundify categorizes the sources of sound in order to accurately match sound effects with videos. Then, intervals are established by comparing the labels of the effects with each frame and identifying repeated matches that exceed a predefined threshold. The mix portion of the process segments the effects into 1 sec intervals, which are subsequently merged together using cross-fading [18]. Soundify synchronizes the sound clips with the video and converts them into spatial audio by altering the panning and volume based on CLIP's activation maps [30].

9.2.4 Audio Generation Models

Audio generation models mostly concentrate on the creation of audio content, such as the production of speech using artificially generated voices that mimic those of humans. Text-to-speech and speech-to-speech models have the potential to be utilized in a wide range of applications, including digital assistants, customer services, audiobook and training narration, and accessibility tools. Businesses operating in the music industry can utilize high-fidelity audio generation to produce tailor-made soundtracks for marketing, films, or video games, resulting in substantial reductions in both cost and time compared to conventional music production methods [5]. We will look at two models that accept text as input and output audio: VALL-E and MusicLM.

9.2.4.1 MusicLM

MusicLM approaches conditional music generation as a hierarchical sequence-to-sequence modeling task, resulting in the production of music at a frequency of 24 kHz that remains consistent over several minutes [31]. MusicLM assists musicians in their creative process by providing inspiration and facilitating the development of sophisticated musical compositions [5]. MusicLM employs three models to extract audio representations that are utilized for the purpose of generating music through conditional autoregressive methods. The acoustic tokens generated by SoundStream's self-supervised audio representations are utilized for high-fidelity synthesis. Additionally, the semantic tokens produced by w2v-BERT are employed to support long-term coherent generation. The conditioning is represented via the MuLan music embedding during training. Each of these three models is pretrained separately and then locked in place so that they may give distinct audio and text representations for sequence-to-sequence modeling. MusicLM produces exceptional music by utilizing a textual description, hence expanding the range of resources that aid humans in creative musical endeavors [31].

9.2.4.2 VALL-E

Microsoft's VALL-E enhances user experience by providing realistic voice modeling, resulting in a more personalized and engaging interaction. By providing specific genres or melodies via prompts, unique pieces of music can be generated that respect the original intent [5]. VALL-E is a language model that uses a cascaded technique for text to voice (TTS) using audio codec codes as intermediate representations. VALL-E is pretrained with 60,000 hours of speech data and provides the in-context learning capability in zero-shot scenarios. In addition, VALL-E has the ability to accurately reproduce the acoustic environment and capture the speaker's emotions during synthesis. It also offers a wide range of outputs using various sampling-based decoding techniques. VALL-E possesses in-context learning capabilities and has the ability to generate high-quality personalized speech using only a 3 sec recording of an unfamiliar speaker as an acoustic stimulus. VALL-E has the capability to generate various outputs using the same input text while preserving the acoustic environment and the emotional expression of the speaker in the acoustic prompt [32].

9.2.5 Code Generation Models

Code generation models have the potential to transform the way developers work and write software. Automated code generation, idea conversion into executable scripts, auto-completion features, created unit tests, duplicate code identification, and bug fixes are all available to programmers utilizing tools like GitHub Copilot. These auto-mated possibilities allow developers to work on more complex projects and problem-solv-ing, which improves output and the overall quality of the finished product, reduces time to market, facilitates quick prototyping, and encourages ongoing innovation for the business and product [5]. Even though there are a lot of models that may be applied to natural language writing, it is crucial to keep in mind that different types of text have different syntax, especially when it comes to computer code [18]. We will look at two models that accept text as input and output code: Codex and AlphaCode.

9.2.5.1 Codex

Codex is an AI system, created by OpenAI, that translates language into codes. With the help of this all-purpose programming technique, programmers can break down complex problems into smaller, simpler ones that can be mapped to pre-existing code libraries, APIs, or processes [18]. OpenAI trained GPT-3 using code from GitHub to develop the Codex model. The resulting model efficiently translates com-mon languages into computer code, demonstrating strong performance on a dataset of human-written issues of varying difficulty [33]. Codex received training using a dataset of 179 gigabytes consisting exclusively of distinct Python scripts that were all under 1 megabyte. The files were obtained from GitHub's open-source software repository in May 2020. Codex's fine-tuning is built upon the powerful natural lan-guage processing capabilities of GPT-3 [18].

9.2.5.2 AlphaCode

The language model AlphaCode was created for code creation and is capable of handling complex logical processes. The key to its effectiveness lies in the combi-nation of several components, such as a substantial and efficient transformer-based framework, a comprehensive model sampling approach, and an extensive dataset for both training and evaluation purposes. The dataset utilized by AlphaCode for train-ing consists of code extracted from GitHub projects, amounting to a total of 715.1 GB. This dataset is notably larger than the dataset employed by Codex for pretrain-ing. The model is fine-tuned using a dataset obtained from the Codeforces platform. Codeforces, a coding competition platform, provides a valuable dataset for validating models and improving performance [18].

9.2.6 OTHER MODELS

The applications of general AI go beyond the specified categories and domains, affecting various other specific areas [5]. For example, there are several text-to-3D-models tools available, such as DreamFusion, Nvidia GET3D, Magic3D, and Point-E. These tools are capable of generating very realistic and intricate 3D models which may be used in various industries, like product design, architecture, virtual reality, and game development. Image-to-text models are the opposite of text-to-image synthesis models like Flamingo and VisualGPT. Text-to-molecules models, such as AlphaFold and OpenBioML, produce functional protein structures and create new molecules by producing valid and innovative molecular structures. These models provide valuable support to researchers in drug discovery and bioengineering [5, 18].

9.3 GENERATIVE AI APPLICATIONS

AI has attracted global interest in recent years since the launch of OpenAI's application called "Chat Generative Pre-trained Transformer," or ChatGPT, in late 2022 [34]. The development of ChatGPT has led to a significant debate over the integration of artificial intelligence in several fields, including academics, business, and society as a whole. While AI has been employed in various domains for many years, the advent of generative AI programs like ChatGPT, Jasper, or DALL-E is seen as a significant advancement in the progress of AI technology. This is mostly because of their user-friendly nature, intuitive interface, and impressive performance. By 2023, there are multiple GAI systems that specialize in various activities, such as text generation, image manipulation, video editing, audio production, and coding [16]. This section presents the usability of GAI applications in several domains.

9.3.1 BUSINESS

Generative AI, like ChatGPT, can provide various benefits to enterprises across multiple domains, including marketing, sales, operations, IT/engineering, risk and legal, human resources, accounting and finance, as well as utility/employee optimization. For instance, it can operate as a chatbot to provide customer support, act as a virtual assistant to assist consumers in accomplishing specified tasks, perform accounting and human resource functions, and develop advertisements or marketing concepts. ChatGPT may serve as both an internal and external collaborator for a wide range of company projects or campaigns. Therefore, the possible business applications of ChatGPT are limitless [34].

9.3.2 SOFTWARE ENGINEERING

Artificial intelligence has been significantly transforming the software industry for a considerable period of time. Software developers can enhance their efficiency by automating repetitive processes, optimizing the debugging process, streamlining testing, and accessing several additional functions. The enhanced availability of GenAI products offers software engineers greater benefits in their daily lives [35]. GAI is expected to have a significant impact on the field of software engineering through the use of automated code creation, documentation, issue identification, and performance optimization. According to internal sources at Google, it has been reported that ChatGPT has the potential to successfully pass interviews for an entry-level software engineer position. By utilizing cloud-based platforms and internet services, the sector has the potential to be more easily accessible to individuals who lack technological expertise. The emergence of GAI in software engineering has the potential to move developers' focus from coding to making strategic decisions. Some experts compare this transition to the huge impact of the cloud or DevOps [16].

9.3.3 EDUCATION

ChatGPT has caused significant disruptions and brought about significant shifts in every aspect of education. ChatGPT works as a valuable assistant in educational attempts, both for learning and teaching purposes. ChatGPT can aid students in a range of tasks, including information retrieval, addressing queries related to certain subjects, and improving writing proficiency across multiple languages. ChatGPT can aid teachers in creating teaching plans; developing teaching materials, such as scripts, slides, and quizzes; evaluating and grading assignments; and offering feedback to students. ChatGPT, being developed on language models (LLMs), has the capability to generate instructional material, customize learning experiences, and enhance student involvement. This can lead to increased efficiency and effectiveness in delivering education. Within the realm of academic research, ChatGPT may provide valuable support in various areas, such as problem understanding, research methodology development, data gathering and analysis, as well as evaluating and providing feedback on written work and its structure. While ChatGPT proves to be beneficial in educational settings, there are worries regarding its potential misuse by students for cheating in exams or engaging in plagiarism when writing essays. The adverse consequences include the disruption of educational norms, a breakdown of students' learning process, and the endangerment of academic integrity [34]. Both professors and administrators are actively seeking solutions to address the repercussions caused by the influence of ChatGPT. Some are fearful and angry at this disruption, while others are already using it in class, asking their students to use it in new and creative ways [22]. For example, the University of Cambridge has declared that including work produced by AI platforms like ChatGPT will be classified as academic misconduct. Oxford University maintains a receptive stance toward ChatGPT and regards it as a valuable resource for both teachers and learners. A range of AI content detection services is offered by CatchGPT, AI Writing Check, Content at Scale, Copyleaks, GPTZero, OpenAI's AI Text Classifier, and ZeroGPT [34].

9.3.4 HEALTHCARE

Since the middle of the 20th century, medicine has been seen as one of the most promising application areas of AI [21]. Generative AI and ChatGPT can make a big difference in the field of healthcare. With the help of LLMs, generative AIs like ChatGPT could change many parts of the healthcare business [34]. For example, GAI can aid in speeding up the discovery of new drugs, supporting clinical trials, caring for patients, estimating health state through wearable tech, and performing automatic robotic surgery. In the field of medical AI, automated medical picture diagnosis is the most promising area, especially in image-based diagnosis fields like radiology, ophthalmology, and dermatology [21]. Encrypted GAI robots can be trusted friends who listen when people want to talk about their feelings or mental health issues. Giving doctors the tools to look at huge amounts of medical data, make more accurate diagnoses, and make choices based on that data could free up a lot of time for them [16]. AI can also help doctors give patients the most up-to-date medical information from journals, textbooks, and clinical practices. It can also help them get useful information from big groups of patients so they can get real-time health risk alerts and predictions about how their health will progress [21]. GenAI can take X-rays and CT scans and turn them into more accurate pictures. This could help doctors figure out what is wrong. Medical professionals can learn more about a patient's internal parts with the help of GANs (generative adversarial networks). This method can help a lot because it enables doctors to find life-threatening diseases like cancer at an early stage [36].

9.3.5 MEDIA AND CONTENT CREATION

GenAI is significantly influencing the media industry and transforming both the creation and consumption of content. GAI models have the capability to generate a diverse range of information, such as text, photographs, videos, and audio, leading to faster and more efficient production while reducing costs [36]. For instance, the marketing sector is adopting generative AI to create synthetic and customized adverts for prospective consumers. Synthetic advertising is created and modified using artificial and automated methods of generating data [34]. Generative AI has the capability to personalize content for individual users, resulting in higher user engagement and retention. Virtual assistants can provide assistance in various areas, such as exploring novel information, arranging one's schedule, and performing voice-activated searches [36]. Generative AI has significantly transformed the journalism profession. News robots like Quill and Xiaomingbot have been employed in the field of news production. Their primary focus is on generating news content that heavily relies on data analytics and follows a somewhat strict template structure. Therefore, generative AI has the ability to create news articles that are more intricate, incorporating both text and videos. The field of art production is currently undergoing significant transformations due to the emergence of generative AI models like ChatGPT, DALLE-2, and Midjourney [34].

9.3.6 FINANCIAL SERVICES

The future potential of GAI resides in its ability to influence investment management. Through the utilization of advanced algorithms to analyze extensive financial data, GAI has the capability to detect patterns, forecast market trends, and discover investment opportunities. Ultimately, this would improve investment strategies, maximize returns, streamline due diligence, and boost portfolio performance [16]. By employing GenAI to analyze client expenditure patterns and detect potential concerns, banks and other financial institutions can get novel insights into customer behavior and become cognizant of potential issues [36]. Furthermore, GAI has the potential to assist in the identification and prevention of fraudulent activities, as well as perform customer-oriented functions, including providing tailored financial guidance in the field of wealth management. Evidently, there is already an obvious change occurring in the financial landscape, marked by the emergence of innovative tools like PitchBook's IPO prediction software, Finchat, which is sometimes referred to as the "ChatGPT of finance," and the introduction of Bloomberg's own language model boasting an impressive 50-billion parameter capacity. Another instance is GPTQuant, a conversational AI chatbot designed for the purpose of designing and assessing investment plans [16].

9.4 CHALLENGES WITH GENERATIVE AI

Within the realm of AI, ethical problems concern the moral responsibilities and obligations of both the AI application and its developers. Here, we examine the challenges associated with implementing generative artificial intelligence in practical scenarios. These challenges include misuse, privacy and security, bias, harmful or inappropriate content, overreliance, data quality and accessibility, copyright, regulatory frameworks and policy development, and incorrect outputs.

9.4.1 MISUSE

Generative AI misuse involves any intended utilization that may lead to harmful, unethical, or unsuitable consequences [34]. Generative AI can be utilized to produce harmful material, misinformation, and present threats to a cohesive society. Protecting AI systems from malicious misuse and maintaining the integrity of created content are crucial problems [37]. Education is a major field that is vulnerable to misuse. Considering the capability of generative AI like ChatGPT to produce prompt and excellent responses in a matter of seconds, students lacking motivation may not invest sufficient time and effort into completing their tasks and essays. Hence, assessing the novelty of students' work may prove a difficulty. Another instance of improper use is the practice of participating in cheating during examinations. Permitting students to access digital devices during examinations would enable them to employ ChatGPT as a resource to assist them in answering questions. In order to address these difficulties, one possible solution is to employ AI-powered content detectors like Turnitin [34]. Another kind of misuse is the creation of fake images with the intention of causing harm, disseminating false information, or violating copyright and intellectual property rights [27]. The widespread use of GAI models can lead to the dissemination of false information and the manipulation of the media and politics, or the perpetration of fraud against persons and corporations. Generative AI researchers aim to establish criteria for enhancing safety and promoting responsible utilization. Nevertheless, it is possible for applications to be deceived into circumventing filters and protections of GAI models. This can be accomplished, for example, by injecting malicious prompts that result in misaligned outputs of generative AI applications [5].

9.4.2 PRIVACY AND SECURITY

Data privacy and security represent significant challenges for generative AI models like ChatGPT. Data security involves the measures used to protect information from unauthorized access, modification, or theft. *Privacy* refers to confidential personal information that individuals wish to keep undisclosed to others. During the development phase of ChatGPT, a significant amount of personal and sensitive data was used for training reasons, providing a potential risk to privacy. System issues in ChatGPT have resulted in the unintended visibility of chat logs for certain users. Both individual users and large organizations or government agencies encounter information privacy and security concerns [34]. Ensuring explicit consent and safeguarding personal data are crucial for the ethical implementation of these technologies [27]. In order to deal with concerns related to privacy and security, it is crucial for users to exercise caution when engaging with ChatGPT, to prevent the accidental sharing of sensitive personal or private organizational information. Large digital corporations should take steps to increase user awareness on ethical issues, including privacy and security. Simultaneously, it is imperative to establish legislation and procedures that safeguard the privacy and security of information [34].

9.4.3 BIAS

In the field of artificial intelligence, *bias* refers to the ability of AI-generated responses or suggestions to exhibit inappropriate preference or discrimination toward specific individuals or groups. Various types of biases are occasionally detected in the material produced by language models, which may be a result of the training data. Considering cultural sensitivities in different nations is crucial to prevent biases, given ChatGPT's widespread international presence [34]. Societal biases are present in all aspects of human-generated content. Training deep learning models with biased data can magnify human biases, reproduce poisonous language, or propagate stereotypes related to gender, sexual orientation, political affiliation, or religion [19]. Therefore, it is crucial to have training data that is representative, comprehensive, and diverse in order to ensure fairness and prevent biases. It is important to subject generative AI applications to testing and evaluation by a variety of users and subject experts. Moreover, enhancing the transparency and comprehensibility of generative AI can aid in the identification and detection of biases, hence enabling the implementation of suitable remedial actions [34]. The issue of bias and justice in AI is receiving more and more attention in academic literature, yet it is still an unresolved and ongoing research challenge. At both the system and application levels, it is possible to integrate mitigation techniques to specifically target biases that are ingrained in deep learning models. This can result in the generation of outputs that are more varied and diversified [19]. In order to protect users and maintain the quality and reputation of a company, it is important to create strategies that can prevent, identify, and minimize biases [5].

9.4.4 HARMFUL OR INAPPROPRIATE CONTENT

Generative AI can produce harmful or improper content, such as violent material, offensive language, discriminatory content, and pornography. Despite OpenAI's implementation of a content policy for ChatGPT, the presence of harmful or improper content can still occur as a result of algorithmic constraints or unauthorized modifications. *Toxicity* is the term used to describe the language models' capacity to comprehend or produce damaging or offensive content. The presence of toxicity can have detrimental effects on society, causing harm and disrupting the harmony within the community. Hence, it is crucial to ensure that any harmful or disagreeable material is not present in the training data and swiftly removed if identified. Similarly, the training data must be free of any pornographic, sexual, or erotic content. Implementing rules, regulations, and governance is crucial to prevent the dissemination of any offensive content to users [34].

9.4.5 OVERRELIANCE

The perceived ease and success of ChatGPT may lead to overreliance among its users, causing them to place strong trust in the answers delivered by ChatGPT. Unlike standard search engines that offer several information sources for users to evaluate and choose from, ChatGPT produces precise responses customized to each request. Utilizing ChatGPT offers the advantage of increasing efficiency by saving time and effort. However, users may acquire a tendency to accept responses without enhancing or checking them. Excessive dependence on generative AI technology might slow the progress of skills, such as creativity, critical thinking, and problem-solving. [34]. Although AI technologies have the potential to enhance human capabilities, they are incapable of replacing human intuition, empathy, and creativity [37]. Therefore, it is important for users to exercise attention when relying on the responses generated by AI systems and to verify them before accepting them as accurate [34].

9.4.6   DATA QUALITY AND ACCESSIBILITY

Generative AI relies primarily on massive datasets for training and generating answers [37]. The efficacy of generative AI models is essentially governed on the level of quality of the training data. The model's output could reflect factual inaccuracies, conflicting information sources, or biases that exist in the training data [34]. Addressing the lack of data is crucial for ensuring the accuracy and dependability of GAI applications, particularly in the fields of healthcare, schooling, and income management [37]. Generative AI models such as ChatGPT or Stable Diffusion typically require substantial quantities of training data. Having both high-quality and complete, balanced datasets is crucial. It is important to perform data cleansing on the training datasets. However, this process might be excessively costly due to the vast quantity of data involved. Utilizing synthetic training data can serve the dual purpose of enhancing dataset diversity and mitigating sample selection biases [34].

9.4.7   COPYRIGHT

Generative AI models and apps possess the capacity to contravene copyright laws by producing outputs that closely resemble or mimic existing works without receiving permission or offering a reward to the original creators [19]. *Copyright* is a legal concept that safeguards original creative works once they have been documented in an intangible form. Some content supplied by GAI may include original works produced by others that are safeguarded by copyright laws and regulations [34]. Generative AI has the potential to create unauthorized copies of a work, so infringing upon the reproduction right of creators [19]. Consequently, it is necessary to acknowledge authorship to generative AI, which has important consequences for the underlying principles of copyright law and creativity. It is imperative to carefully explore the design and execution of rules, regulations, and laws pertaining to the proper use of generative AI [34].

9.4.8   ENVIRONMENTAL IMPACT

Training large-scale generative AI models requires a lot of computing, which has a big environmental impact. AI training's consumption of energy presents sustainability concerns, particularly for climate action [37]. Finally, because generative AI systems are built around large neural networks, they consume a lot of electricity and have a large negative carbon footprint. For instance, training a generative AI model like GPT-3 produced 552 t CO_2, equivalent to the annual CO_2 emissions of several hundreds of families [19]. OpenAI's power consumption in January 2023 was similar to 175,000 Danish families' annual consumption, and AI might affect millions of people's electricity use. The growing demand for data centers due to generative AI and other AI applications is straining local energy networks and harming new housing constructions. Climate change and increasingly frequent heat waves are expected to increase demand for power systems and data centers, potentially disrupting operations [21]. Due to this, AI researchers are working to make AI algorithm development and deployment more carbon-friendly, by improving training algorithms, compressing neural network architectures, and optimizing hardware [19]. AI technology's environmental impact must be reduced by developing energy-efficient AI algorithms and implementing sustainable behavior [37]. Generally, AI's environmental consequences depend on its design, implementation, and management. Like any technical breakthrough, it is important to examine environmental impacts and minimize bad effects while maximizing advantages [21].

9.4.9 REGULATORY FRAMEWORKS AND POLICY DEVELOPMENT

The rapid advancement of AI technology frequently exceeds the progress made in constructing regulatory frameworks and rules. Establishing an optimal balance between promoting innovation and guaranteeing appropriate utilization of artificial intelligence requires the implementation of strong rules. Collaboration between governments and international organizations is crucial for the development of ethical norms, standards, and laws that will govern the implementation of AI technology [37]. Improving the design and implementation of AI governance can be achieved by advocating for transparency and explainability of AI systems, as well as encouraging collaboration between technological giants and the government [34].

9.4.10 INCORRECT OUTPUTS

Generative AI models may generate output with errors. This is because machine learning models use probabilistic algorithms to make inferences. For instance, generative AI algorithms generate the most likely prompt response, not the right one [19]. Studies have shown GAI-based picture generators that produce human anatomical errors. These errors show that GAI models need improvement before they can be employed for unsupervised production [5]. LLMs' generative model generates text with semantically or syntactically reasonable but nonsensical errors. These models may generate material based on assumptions or biases rather than facts. Additionally, generative AI output, especially LLM output, is rarely verified. Correctness checks can prevent certain outputs in generative AI systems and applications. Generative AI can generate explanations or references that users can verify to address downstream effects of inaccurate results. Such explanations are uncertain and vulnerable to errors, but they may assist users in deciding when to trust generative AI outcomes [19].

REFERENCES

1. Goodfellow, I., Bengio, Y., & Courville, A. (2016). *Deep learning*. MIT Press.
2. Choi, R. Y., Coyner, A. S., Kalpathy-Cramer, J., Chiang, M. F., & Campbell, J. P. (2020). Introduction to machine learning, neural networks, and deep learning. *Translational Vision Science & Technology*, 9(2), 14–14.
3. Varghese, J., & Chapiro, J. (2024). ChatGPT: The transformative influence of generative AI on science and healthcare. *Journal of Hepatology*, 80(6), 977–980.
4. Kalyan, K. S. (2023). A survey of GPT-3 family large language models including ChatGPT and GPT-4. *Natural Language Processing Journal*, 100048.
5. Banh, L., & Strobel, G. (2023). Generative artificial intelligence. *Electronic Markets*, 33(1), 63.
6. Russell, S. J., & Norvig, P. (2016). *Artificial intelligence: A modern approach*. Pearson.
7. Janiesch, C., Zschech, P., & Heinrich, K. (2021). Machine learning and deep learning. *Electronic Markets*, 31(3), 685–695.
8. Guo, Y., Liu, Y., Oerlemans, A., Lao, S., Wu, S., & Lew, M. S. (2016). Deep learning for visual understanding: A review. *Neurocomputing*, 187, 27–48.
9. Gamoura, S. C., Koruca, H. İ., & Urgancı, K. B. (2023, May). Exploring the transition from "contextual AI" to "generative AI" in management: Cases of ChatGPT and DALL-E 2. In *International Symposium on Intelligent Manufacturing and Service Systems* (pp. 368–381). Springer Nature.
10. Aydin, Ö., & Karaarslan, E. (2023). Is ChatGPT leading generative AI? What is beyond expectations? *Academic Platform Journal of Engineering and Smart Systems*, 11(3), 118–134.
11. Rossi, S., Rossi, M., Mukkamala, R. R., Thatcher, J. B., & Dwivedi, Y. K. (2024). Augmenting research methods with foundation models and generative AI. *International Journal of Information Management*, 102749.
12. Liao, W., Lu, X., Fei, Y., Gu, Y., & Huang, Y. (2024). Generative AI design for building structures. *Automation in Construction*, 157, 105187.
13. OpenAI, R. (2023). Gpt-4 technical report. *View in Article*, 2(5). arxiv 2303.08774.
14. Liu, V., & Chilton, L. B. (2022, April). Design guidelines for prompt engineering text-to-image generative models. In *Proceedings of the 2022 CHI Conference on Human Factors in Computing Systems* (pp. 1–23). Association for Computing Machinery, New York, NY, United States.
15. Bandi, A., Adapa, P. V. S. R., & Kuchi, Y. E. V. P. K. (2023). The power of generative ai: A review of requirements, models, input–output formats, evaluation metrics, and challenges. *Future Internet*, 15(8), 260.
16. Kanbach, D. K., Heiduk, L., Blueher, G., Schreiter, M., & Lahmann, A. (2023). The GenAI is out of the bottle: Generative artificial intelligence from a business model innovation perspective. *Review of Managerial Science*, 1–32.
17. Rios-Campos, C., Viteri, J. D. C. L., Batalla, E. A. P., Castro, J. F. C., Núñez, J. B., Calderón, E. V., . . . Tello, M. Y. P. (2023). Generative artificial intelligence. *South Florida Journal of Development*, 4(6), 2305–2320.
18. Kumar, S., Musharaf, D., Musharaf, S., & Sagar, A. K. (2023, June). A comprehensive review of the latest advancements in large generative AI models. In *International Conference on Advanced Communication and Intelligent Systems* (pp. 90–103). Springer Nature.
19. Feuerriegel, S., Hartmann, J., Janiesch, C., & Zschech, P. (2024). Generative AI. *Business & Information Systems Engineering*, 66(1), 111–126.
20. Baidoo-Anu, D., & Ansah, L. O. (2023). Education in the era of generative artificial intelligence (AI): Understanding the potential benefits of ChatGPT in promoting teaching and learning. *Journal of AI*, 7(1), 52–62.
21. Mannuru, N. R., Shahriar, S., Teel, Z. A., Wang, T., Lund, B. D., Tijani, S., . . . Vaidya, P. (2023). Artificial intelligence in developing countries: The impact of generative artificial intelligence (AI) technologies for development. *Information Development*, https://doi.org/10.1177/02666669231200628.

22. Sætra, H. S. (2023). Generative AI: Here to stay, but for good? *Technology in Society*, 75, 102372.
23. OpenAI. (2024). *Introducing improvements to the fine-tuning API and expanding our custom models program*. https://openai.com/blog.
24. Barreto, F., Moharkar, L., Shirodkar, M., Sarode, V., Gonsalves, S., & Johns, A. (2023, February). Generative artificial intelligence: Opportunities and challenges of large language models. In *International Conference on Intelligent Computing and Networking* (pp. 545–553). Springer Nature.
25. Espejel, J. L., Ettifouri, E. H., Alassan, M. S. Y., Chouham, E. M., & Dahhane, W. (2023). GPT-3.5, GPT-4, or BARD? Evaluating LLMs reasoning ability in zero-shot setting and performance boosting through prompts. *Natural Language Processing Journal*, 5, 100032.
26. Schick, T., Dwivedi-Yu, J., Jiang, Z., Petroni, F., Lewis, P., Izacard, G., You, Q., Nalmpantis, C., Grave, E. & Riedel, S. (2022). Peer: A collaborative language model. arXiv preprint arXiv:2208.11663.
27. Suryadevara, C. K. (2020). Generating free images with OpenAI's generative models. *International Journal of Innovations in Engineering Research and Technology*, 7(3), 49–56.
28. Turchi, T., Carta, S., Ambrosini, L., & Malizia, A. (2023, May). Human-AI co-creation: Evaluating the impact of large-scale text-to-image generative models on the creative process. In *International Symposium on End User Development* (pp. 35–51). Springer Nature.
29. Villegas, R., Babaeizadeh, M., Kindermans, P. J., Moraldo, H., Zhang, H., Saffar, M. T., Castro, S., Kunze, J. & Erhan, D. (2022, September). Phenaki: Variable length video generation from open domain textual descriptions. In *International Conference on Learning Representations*.
30. Lin, D. C. E., Germanidis, A., Valenzuela, C., Shi, Y., & Martelaro, N. (2023, October). Soundify: Matching sound effects to video. In *Proceedings of the 36th Annual ACM Symposium on User Interface Software and Technology* (pp. 1–13). Association for Computing Machinery, New York, NY, United States.
31. Agostinelli, A., Denk, T. I., Borsos, Z., Engel, J., Verzetti, M., Caillon, A., Huang, Q., Jansen, A., Roberts, A., Tagliasacchi, M. & Sharifi, M. (2023). Musiclm: Generating music from text. arXiv preprint arXiv:2301.11325.
32. Wang, C., Chen, S., Wu, Y., Zhang, Z., Zhou, L., Liu, S., Chen, Z., Liu, Y., Wang, H., Li, J. & He, L. (2023). Neural codec language models are zero-shot text to speech synthesizers. arXiv preprint arXiv:2301.02111.
33. Brennan, R. W., & Lesage, J. (2022, September). Exploring the implications of OpenAI codex on education for industry 4.0. In *International Workshop on Service Orientation in Holonic and Multi-Agent Manufacturing* (pp. 254–266). Springer International Publishing.
34. Fui-Hoon Nah, F., Zheng, R., Cai, J., Siau, K., & Chen, L. (2023). Generative AI and ChatGPT: Applications, challenges, and AI-human collaboration. *Journal of Information Technology Case and Application Research*, 25(3), 277–304.
35. Petrovska, O., Clift, L., Moller, F., & Pearsall, R. (2024, January). Incorporating generative AI into software development education. In *Proceedings of the 8th Conference on Computing Education Practice* (pp. 37–40). Association for Computing Machinery, New York, NY, United States.
36. Mandapuram, M., Gutlapalli, S. S., Bodepudi, A., & Reddy, M. (2018). Investigating the prospects of generative artificial intelligence. *Asian Journal of Humanity, Art and Literature*, 5(2), 167–174.
37. Rane, N. (2023). Roles and challenges of ChatGPT and similar generative artificial intelligence for achieving the sustainable development goals (SDGs). Available at SSRN http://dx.doi.org/10.2139/ssrn.4603244.

10 Ethical Dimensions of Artificial Intelligence Balancing Innovation and Responsibility

Mehmet Milli

10.1 ARTIFICIAL INTELLIGENCE

Artificial intelligence (AI) is generally known to everyone as a discipline designed for computer systems to have human-like intelligence. However, from its first emergence to the present day, it has developed and naturally changed, just like other disciplines. Therefore, although it is possible to see different definitions in different sources since its first appearance, the basic principle and purpose are always the same. The main purpose of the AI workspace is to enable computers to have the ability to solve complex problems and make decisions. AI aims to achieve this goal by using techniques such as data analysis, pattern recognition, pattern detection, machine learning, and natural language processing. Despite a long history of research and discussion, there is still no standard definition of AI. Some of the definitions from its emergence to the present are given in the following.

- According to Turing, to say that a machine is intelligent, its behavior must become indistinguishable from that of a human being [1]. This idea forms the basis of a test known today as the "Turing test." If a computer program passes the Turing test, it means that the program can exhibit intelligent behavior and be considered artificial intelligence.
- A very frequently used definition in literature is made by John McCarthy. According to the definition he made in his 2007 article "What Is Artificial Intelligence?" he explains that "intelligence" means the computational part of the ability to achieve goals in the world [2]. In this article, the concept of AI is defined as the science and engineering of making intelligent machines.
- Today's definition of modern AI is getting more and more specialized. In recent years, we frequently see the term "agent" in the definition of AI in many studies. One of these resources is S. Russell and P. Norvig's book *Artificial Intelligence: A Modern Approach*, in which the fundamentals and different approaches of AI are discussed in detail [3]. In this resource, "intelligence" is explained as meaning the ability to choose an action that is expected to maximize a performance measure. He defines AI as the study

DOI: 10.1201/9781003501152-10

of intelligent agents. The term "agent" in this definition means a software system that senses its environment through sensors and acts according to this environment through actuators.

- In a more recent study, *AI* was defined by M. Ergen as another powerful technological wave that enables a machine to perform cognitive functions, such as perceiving, reasoning, learning, and interacting, just like humans [4]. In the study, it is underlined that the most important feature of the human being is intelligence, and how much the concept of intelligence endowed with human beings can be realized in an artificial environment is discussed and revealed in all its aspects.

Although these definitions are very common definitions of AI. There are numerous definitions used in the literature; however, when we examine the preceding definitions, it is seen that there are no ethical and moral terms in any of them, and no restrictions are imposed on this issue. Apart from these, no matter what definition of *AI* we consider in the literature, the ethical problems brought by artificial intelligence are not mentioned. Currently, ethical concerns brought about by AI are being voiced loudly in many countries. Many companies and institutions not only have to reap the benefits of AI but also must deal with the concerns of the problems it brings.

For companies or institutions to cope with the ethical problems brought by AI, what AI can do, which applications will be accepted as AI, and where AI violates personal rights and privacy should be tied to clear rules. However, the extent of AI-related decisions and limitations is not a decision that only companies can make. While defining ethical AI and determining its boundaries, a consortium including field experts, lawyers, company representatives, end user representatives, policymakers, and various industry representatives should be formed.

Apart from this, the ethical problems brought by the field of AI work today have become a global problem rather than a problem of a particular state, race, or region. Therefore, the presence of experts from different regions and countries in the consortium to be formed ensures that the definition and limitations of ethical AI can be more general and acceptable. Besides, there are many different nationalities and races in the world. Therefore, every nation has its own values and moral limits. Thus, the presence of stakeholders from different nationalities in the consortium to be formed will increase the inclusiveness and acceptability of ethical AI worldwide.

In this study, we will try to list the possible risks that AI applications, which we see frequently in every area of life, may cause in the near future. We will also discuss how companies or institutions should avoid these potential risks, and how they should take precautions while protecting the personal rights and privacy of users. Then we will try to explain the necessity of standardization that will be accepted worldwide for the possible problems of AI in the future. The necessity of establishing a multinational, versatile, and comprehensive consortium will be discussed to achieve this standardization. The structure of the consortium to be formed will be explained as well during the determination of ethical AI rules, which have become a global problem. In this E-AI consortium, which is planned to be formed, the limits of everyone's duties to ensure standardization will be discussed.

10.2   HISTORY OF ARTIFICIAL INTELLIGENCE

Artificial intelligence (AI) is a concept that has been widely used today and is encountered in almost every field of technology. AI, which has become an indispensable part of many advanced systems, is expected to continue to increase its popularity in the future. As a product of studies aimed at imitating some aspects of human intelligence and improving the thinking and learning abilities of machines, AI has had a very interesting journey when considered from a historical perspective. The foundations of AI thought are as old as human history, although it was not called that way back then. These foundations were laid with automatons, hand tools, and mechanical devices that people have made using their imaginations for centuries. Since ancient times, people have designed lifelike beings and mechanical devices with their own creativity, with their thoughts and dreams. However, the concept of AI in the modern sense only emerged in the middle of the 20th century.

In the period after the first quarter of the 1900s, AI gained a different dimension, and in these studies, the necessary maturation was achieved to reveal the first examples of today's AI concept [5–8]. Although AI-like studies date back much earlier, the first study that fully meets today's AI concept was made by Alan Turing. The concept of using computers to simulate intelligent behavior and critical thinking was first described by Alan Turing in 1950. For this reason, Alan Turing is known as the father of AI in the computer world. In these studies, Turing brought up the issue of computers that would imitate the human thinking process before fully understanding how the human mind works [9, 10]. Turing made a proposition that goes further in the field of AI. In his article, he suggested that a model is possible in which a computer is not only programmed to perform a particular task but can also acquire the ability to learn. Accordingly, a machine can acquire new knowledge and improve itself through experiences and data analysis. Alan Turing's thoughts on AI and the Turing test have been an important milestone in the field of AI and are still considered an interesting subject that forms the basis of AI research.

Immediately after Alan Turing introduced the heuristic test to the academic community, studies in the field of AI gained momentum, and in a conference held in 1956, the ability of machines to imitate human intelligence was referred to as AI for the first time. After this definition, which would be accepted by everyone in the future, a few years later, in 1959, "machine learning," a concept that is very often referred to with AI today, was put forward by Arthur Samuel [11]. *Machine learning* is the mechanism that discovers some patterns in past observations and uses these patterns to help improve the predictive performance of the results of subsequent actions. With this study, it has been proven that machines, which have gained a different dimension from machine learning, can gain the ability to learn, just like humans, by experiencing potential scenarios that cannot be predicted before, apart from predefined possibilities. Thus, it paved the way for machines to gain the ability to learn, which is the most important feature of human intelligence, and to use what they have learned in the next decision-making mechanism.

Deep learning, another concept that will be used frequently in the future and will become a valuable scientific term, was introduced in 1965 by Alexey Grigorevich Ivakhnenko and Valentin Grigorievich Lapa [12]. In this study, the researchers laid the foundations of today's deep learning operating principle, which simulates the neural networks of the human brain by running several sensors on top of each other. In this study, which can be considered the first neural network and the first ANN, the system is trained using the group method of data handling [13]. This training method tries to find the best network structure instead of finding the optimization of the parameters. Although the foundations of deep learning were laid in these years, large-scale deep network pieces of training that could not be realized in this period limited the development of deep learning due to limitations in data and computational power. For this reason, the deep learning method could not go beyond theoretical studies at that time. Deep learning methods could not be a common method used in practice until the 1990s, when the data processing capacity of machines increased, and other machine learning techniques had to be preferred instead.

In the 1970s and 1980s, AI research entered a period of crisis because of some fundamental limitations and expectations. During this period, there was a process called the "AI winter," and funding and interest in AI declined drastically. However, in the late 1980s and early 1990s, the field was revived with the emergence of new approaches to artificial intelligence. In particular, the development of technologies such as expert systems, genetic algorithms, and artificial neural networks has provided a second birth of artificial intelligence. Although these years were the stagnation period of AIs, there were a few noteworthy studies that entered the literature. One of these studies was described in the article "Inferring the meaning of direct perception" published by Geoffrey E. Hinton in 1980 [14]. In this study, he introduced *backpropagation*, which is presented as an alternative to forward propagation used by the first ANNs and is a self-optimizing mechanism of ANNs without human intervention. With this mechanism, it became possible to systematically adjust the weights of ANN nodes in multiple layers. Another important study in the hibernation periods of AI was conducted in 1989 by Y. Le Cun et al. [15]. In this study, the researchers introduced convolutional neural networks (CNN), which will be used in many image processing studies in the future. A CNN is a deep learning algorithm that can take an input image, assign weight and bias to various objects in the image, and make objects distinguishable from each other, as well as infer the relationships between objects. Today, CNNs are frequently used for image and video recognition, recommendation systems, analysis of medical images, natural language processing, and classification of visual elements.

In the early 1990s, with the increase in the data processing capacity of machines, significant developments began to occur in AI and sub-working fields (machine learning, deep learning, artificial neural networks, etc.). The developments in the field of transistors and hardware that took place in these years accelerated the field of AI by performing data calculations that seemed impossible in the recent past. One of these studies was published in 1992 by Bernhard E. Boser et al., "Support Vector Machines (SVM)," which he introduced [16]. SVM is a machine learning method that can be used in regression analysis and the classification of data effectively for nonlinear problems. Today, the usage areas of SVMs are quite wide. Many fields of study, such as face detection and recognition, text classification, image processing, bioinformatics, and handwriting recognition, can be given as examples of areas where it is widely used. In addition to all these, it is used in generalized predictive control applications in many engineering fields.

Another important work in these years when AI applications gained momentum is long short-term memory (LSTM), which was put forward by Hochreiter and Jurgen Schmidhuber in 1997 [17]. LSTM, which was introduced by the researchers, is a gradient-based approach, and it is a frequently used method, especially in speech-to-text translations. By the 2000s, there was a tremendous increase in data expected to be processed by machines. With the incredible increase in data in these years, the term *big data* emerged as a new concept in literature. This time, it did not seem like it would be enough to increase the data processing capacity of the machines for AI projects and studies such as the hibernation in the 1970s to continue. Therefore, with the new trend brought by the big data concept, data processing techniques and perspectives have changed. An example of these new perspectives in data processing is dividing large chunks of data into small clusters and trying to understand their trends and behavior.

Some of the studies carried out in the field of AI after the 2000s are as follows: The deep belief network (DBN) method, developed by Geoffrey Hinton et al. in 2006, is an important step in the field of artificial intelligence [18]. The deep belief network is a neural network architecture mostly inspired by the field of probability and graph models. The DBN consists of several layers and is often used for tasks such as classification, feature learning, and pattern recognition. This method offers higher performance and better feature learning capability on certain types of datasets. One of the most important features of DBN is its ability to automatically learn the representation of complex features. This can extract high-level features within datasets as a combination of low-level features, thus providing a better feature representation. This is very useful for better performance in classification and recognition tasks. DBN is mostly used for classification, neural network training, and many other AI tasks. Hinton et al.'s work was carried out using Boltzmann machines containing a particular type of nodes called "hidden variables" to speed up the training of the DBN and achieve better results. The development of DBN marked a milestone in artificial intelligence, highlighting the importance of deep learning and increasing the popularity of deep neural networks. This work also inspired the development of many advanced techniques and algorithms in the field of deep learning and artificial neural networks in the following years.

Another important work is Krishevsky et al.'s deep neural network model named "AlexNet" in 2012 [19]. AlexNet performed exceptionally well in the ImageNet Large-Scale Visual Recognition competition, making it a major milestone in deep learning. AlexNet is quite large and complex compared to other neural network models at that time. Specifically, it was a neural network large enough to be trained based on fast GPUs at that time. The basis of the model is the convolutional neural network (CNN) architecture. AlexNet has an eight-layer CNN structure containing 60 million parameters. The model was trained using the ImageNet dataset of 1.2 million labeled images. The model includes convolution layers, ReLU (rectified linear unit) activation functions, maximum pooling layers, and fully connected layers. The victory of AlexNet caused an explosion in the field of deep learning and formed the basis of many studies in the fields of computer vision and AI. This event triggered the spread of deep learning and great progress in many areas. The 2012 ImageNet competition marked a major advance in AI's ability to understand visual data and recognize objects.

Apart from these studies, GAN, developed by Ian Goodfellow et al. in 2014, has revolutionized object and image production [20]. Consisting of two competing neural networks, this model is known for its ability to generate realistic images and data. ResNet, developed by Kaiming He and his team in 2015, presented a structure with skip links, which is an important feature in solving the difficulties in training multilayer neural networks [21]. In another study in 2016, the AI-based Go game developed by DeepMind became a major turning point by defeating world champion Lee Sedol [22]. A year later, the Word2Vec model, developed by Mikolov et al. in 2017, helped represent word meanings mathematically by creating word vectors in language processing [23]. AlphaFold, also developed by DeepMind in 2020, is known for its achievements in protein structure prediction. AlphaFold, an important development in the field of bioinformatics, used a deep learning–based approach to solve the protein folding problem [24].

Looking at the historical evolution of AI from a different perspective, AI is a modern account of humanity's centuries-long search for reason. Alan Turing's basic questions and mathematical studies laid the foundations of artificial intelligence, and many scientists, engineers, and developers contributed to these studies in the following years. The rapid development of artificial intelligence in the last 20 years has affected many areas of our lives and will continue to bring great changes in the future. These developments have allowed AI to be spread over a wider area and be used to solve more complex problems. Today, AI technologies take place in many areas of our lives, and it is predicted that they will continue to exist as an important part of humanity in the future. However, in addition to this rapid progress, it is of great importance to be careful about issues such as ethics and safety and to direct developments in a way that will positively affect people's lives.

10.3 ARTIFICIAL INTELLIGENCE APPLICATION AREAS

AI, which has undergone great change and development for many years, is used in many fields today. AI processes the raw data held by companies and various institutions and turns it into useful information for these companies and institutions. This information provided by AI provides companies with benefits, such as new markets, greater profitability, and efficient use of the workforce, which may differ according to the sector. AI can improve human life and contribute to the well-being of society by providing more effective and efficient solutions in many areas. In the beginning, these areas are finance, banking, natural language processing, education and learning, e-commerce, agriculture and livestock, health, security and defense, robotics, social media, and automation-based industrial production.

10.3.1 AUTOMATION-BASED INDUSTRIAL PRODUCTION

AI increases efficiency by positively affecting the decision-making processes of automation systems in industrial areas. It helps companies and institutions reach the market more effectively and increases their profitability levels. Combined with AI industrial robots and automation systems, it can make production processes more efficient and control quality. Furthermore, other contributions of AI to companies are as follows:

- Automates business processes and has automatic decision-making capability
- Increases efficiency and reduces costs
- Increases sales rates
- Increases product or service quality
- Provides optimization of business processes, such as supply chain and logistics
- Improves customer satisfaction, loyalty, and experience
- Provides more efficient and improved workforce allocation
- Helps carry out personalized marketing activities

Although AI differs by industry with its functions and areas of use, there are certain advantages it offers to companies. AI helps companies stay one step ahead in an ever-increasing competitive environment by contributing suggestions and data they provide for the decision processes of companies toward their goals and strategies. It is clear that there will be a greater need for AI applications in the field of automation-based industrial production in the future. Therefore, companies must reach the level of technology that can compete with their competitors if they want to find a place in the market in the future, no matter what business field they are in. The keyword in reaching this technological level is undoubtedly the field of AI work. It will be valuable for companies or institutions to make some of their investments or a certain part of their R&D studies on AI studies, which will provide them with tremendous advantages.

10.3.2 BANKING AND FINANCIAL SERVICES

The banking and finance sector is one of the areas that benefit most from AI applications since there are many parameters in its raw data and decision-making processes. In this sector, AI applications are used to obtain very different outputs on different datasets. For example, AI can more precisely assess credit risk by analyzing customer data. This improves banks' lending processes while providing customers with fairer and more personalized loan offers. In another example that can be given in this area, banks can monitor customer account anomalies and detect potential fraud attempts using AI algorithms. Analyzing fraud trends is important to increase the security of customers, and this helps banks increase their credibility.

In banking and finance areas, AI can analyze customer behavior and offer personalized product and service recommendations. This facilitates customers' access to products that suit their needs and preferences. In this way, customer potential can be increased by raising customer satisfaction to higher levels. AI-powered chatbots can be used in customer service to answer customer questions, meet transaction requests, and direct them to solve their problems. This will again increase customer satisfaction by responding to customers' problems as soon as possible. The applications of AI in the banking and finance sector are constantly evolving and becoming widespread. These technologies play an important role in increasing the efficiency of financial institutions, improving customer satisfaction, and reducing risks. Especially since this area is directly related to money, it is vulnerable to attacks and abuse. Therefore, it should not be forgotten that these applications should be used with more attention to ethical issues, such as security and data privacy, than other areas.

10.3.3   EDUCATION

In the field of education, as in all other fields, AI offers several advantages and opportunities to students, teachers, and educational institutions, who are the stakeholders of this field. AI technologies have great potential to improve learning processes, increase student achievement, and adopt a more effective approach to education. Compared to other areas where AI is used, examples of AI applications in the field of education are not sufficient. However, it is clear that AI applications in this field will be encountered more frequently in today's education understanding, where educational materials and education itself have been digitalized recently.

Although it is not very common, it is possible to give some examples of the use of AI in the field of education. For example, AI can offer personalized educational content and lesson plans by analyzing students' individual characteristics, learning styles, and interests. Educating with content suitable for each student's needs and level increases motivation and increases student success. In another example, AI can predict future student success by analyzing data such as student performance, absenteeism rates, and teacher performance. This allows educational institutions and teachers to make better decisions. Apart from all this, AI can guide students in career planning and university selection. It can inform students about suitable career opportunities based on personal skills and interests. It is used to create AI training materials and optimize existing content. This can allow the preparation of more effective and attractive learning materials.

AI provides an important transformation and progress in many aspects of the field of education. Traditional education models are not flexible enough in terms of student diversity and needs, so many students may not have the appropriate learning experience to meet their individual needs. AI is used in education to overcome these challenges and offer a more effective learning process. The use of AI in the field of education has the potential to offer better education opportunities by increasing the quality of education. However, ethical, privacy, and security issues are important considerations in the use of AI in education. In particular, education is one of the most sensitive areas of every nation and state. Therefore, ethical and privacy issues require a degree of sensitivity in the field of education. The responsible use of AI in education must be carried out for the benefit of students and educators.

10.3.4   AGRICULTURE AND LIVESTOCK

The uncontrolled increase in the world population makes it difficult to reach foodstuffs day by day. The fact that both animal and vegetable food cannot be produced as much as before further deepens this problem. Solutions such as managing population growth, preventing food waste, and sustainable use of natural resources can help adopt a more sustainable and equitable approach to food access. In addition to all these, it is thought that the sufficient use of technology in agriculture and animal husbandry and encouraging efficient agricultural practices will contribute to the solution of this problem. Many examples of applications of AI in agriculture and animal husbandry can be given. AI analyzes agricultural data to provide harvest forecasts and productivity-enhancing measures. For example, harvest timing can be better planned using soil moisture data, weather forecasts, and plant growth data collected via sensors. At the same time, suggestions can be made to farmers on issues such as productivity-enhancing practices, irrigation programs, and fertilizer use. In another example, AI develops automated navigation and environmental sensing systems for agricultural robots. These robots can perform different tasks, such as tractors, seeders, and irrigation equipment. Automatic agricultural robots reduce the workload of farmers and save manpower.

To give an example of animal husbandry applications apart from agricultural applications, the first applications that come to mind are undoubtedly animal health and nutrition and feed optimization. AI can analyze sensor data to monitor animal health and detect diseases early. Data such as body temperature, activity level, and feeding habits of animals provide important information about animal health. In this way, the health problems of animals can be detected and treated more quickly. AI can optimize feeding programs by analyzing the nutritional needs of animals. Data on feeding and feed consumption are adjusted according to the needs of the animals, and a healthier growth of the animals can be achieved. AI has great potential to increase productivity and implement sustainable agriculture in the agriculture and livestock sector. These technologies provide farmers with better and more efficient methods, help them use resources more effectively, and help them adopt a sustainable approach to meeting the food needs of society. However, in these applications, just as in other areas, it is very important to carefully apply ethical, security, and data privacy issues to eliminate the social concerns about AI.

10.3.5 MEDICINE AND HEALTH SERVICES

One of the fields of study where AI applications come to life the most is definitely the field of medicine and health services. The periods when diseases were diagnosed and treated using traditional methods in this field are now over. In recent years, AI has become a powerful technology that has revolutionized medicine and health. The importance and impact of artificial intelligence in the health sector are increasing, and it provides benefits in many areas. AI is of great importance in medicine and healthcare and is used in the healthcare industry to meet a range of needs. By using technologies such as AI, data analytics, machine learning, and deep learning, healthcare can become more effective, accessible, and personalized.

Since the use of AI in medicine and health services is quite high, there are many applications designed for different purposes in this field. Although the medical and health applications of AI are very diverse and uncountable, some examples can be given where it is used the most. The most common use of AI applications in the field of medicine and health services is probably in the diagnosis of diseases and treatment planning. AI plays an important role in the diagnosis of diseases by analyzing medical imaging (MRI, CT, X-ray, etc.) and biomedical data. Especially thanks to deep learning algorithms, early diagnosis of diseases such as cancer, cardiovascular diseases, and neurological disorders can be made more sensitively and accurately. Another AI application used in this field is drug development. Molecular structure and interaction analysis can accelerate the development of new drugs. AI-based algorithms can identify potential drug candidates by analyzing large datasets, making this process more cost-effective and time-efficient. AI is also used in areas such as treatment planning and implementation, patient follow-up and management, and patient reporting and documentation.

The applications of AI in medicine and healthcare are creating a major transformation, especially in patient care and treatment. These technologies have the potential to provide more accurate diagnoses, personalized treatments, more effective patient care, and more efficient healthcare. However, the field of medicine and health services is more sensitive than others in terms of ethics and privacy. For this reason, the use of AI in the field of health is an important issue that should also be considered in ethical and safety issues. Data privacy, patient rights, and protection of ethical values are the basic principles of the responsible application of artificial intelligence in medicine and health.

10.3.6 PUBLIC INSTITUTIONS

AI is transforming many industries today, and public institutions are no exception. AI applications, which are increasingly important in public institutions, have become a powerful tool to improve the service quality and efficiency of governments and local administrators. Just as in other industries, AI paves the way for solving complex problems, increasing efficiency, and making better decisions by using technologies such as big data analysis, machine learning, and deep learning in public institutions. It is a fact that AI applications will continue to increase day by day in public spaces as well as in private companies and factories.

When it comes to AI applications in public areas, the first thing that comes to mind is undoubtedly city planning, smart transportation, and smart cities. AI is used in city planning and transportation management. It provides information about traffic congestion and transportation problems by analyzing the flow of traffic. AI-based smart city projects can be developed to reduce the energy consumption and environmental impacts of cities. AI is an important aid in dealing with disasters and emergencies. AI applications in areas such as disaster prediction, emergency planning, and crisis management can be used to provide rapid and effective responses. Another example where AI is used in public institutions would be in social security. AI makes significant contributions to police and security forces. Thanks to technologies such as facial recognition and voice analysis, crime prevention and investigation processes become more effective. AI applications are also used to monitor suspicious activities and predict events such as terrorist attacks.

Apart from these, examples of the use of AI in public areas are the personalization of public services, public health and epidemic control, and citizen participation and service delivery. As a result, AI has great potential to provide efficiency, transparency, and effectiveness in public institutions. However, public areas and services cover and concern the public. Therefore, it is a sensitive issue and open to abuse and attacks. Especially in this area, it is important to carefully manage AI applications on issues such as ethics, security, and data privacy. The potential of artificial intelligence for the development and improvement of public services will continue to increase in the coming years. Therefore, investing in and effective use of AI technologies by governments and local administrators are an important step toward improving the quality of public services.

10.3.7 SOCIAL MEDIA

Social media has become an integral part of the daily lives of millions of people today. Social media continues to grow rapidly as a platform where people connect with friends, share content, and follow news. Having many different social media platforms creates difficulties in terms of keeping and managing the social media environment safe. AI plays an important role in social media platforms and makes these platforms more effective, secure, and personalized. Undoubtedly, the most used area of AI applications in social media is in personalizing content and giving user-oriented recommendations. Social media platforms personalize users' experiences by offering content that matches their interests and behaviors. AI algorithms analyze users' past behavior and tastes, providing users with tailored content and recommendations. In this way, users interact with more interesting and relevant content.

An example of another application where AI is used in this field is in sentiment analysis and meaning extraction. AI analyzes the texts and images shared on social media and works to understand the emotional states and thoughts of users. Sentiment analysis helps social media platforms better understand the content and offer more relevant content based on users' emotional reactions. Another example is determining advertising and marketing strategies. Social media platforms have become an important platform for companies to promote and market their products and services. AI helps manage advertising and marketing strategies more effectively. Targeting based on user behavior and preferences makes ads more effective and engaging.

In addition to all these, it is possible to give many more examples of the use of AI applications in social media. These include trend analysis and news tracking, content moderation and filtering, fraud and fake account detection, and customer service. As can be seen, the use of AI on social media platforms has great potential. However, considering the ethical and data privacy issues of these practices is important in order to gain the trust of users and prevent negative effects. In particular, the difficulty of security and management of social media platforms reveals the necessity of paying more attention to ethics and privacy issues.

10.3.8 NATURAL LANGUAGE PROCESSING

Natural language processing (NLP) is a field of computer science that allows computers to make sense of and process the spoken or written language of humans. It includes the processes of understanding, interpreting, and processing text data, spoken language, and natural language questions. Today, considering that billions of text data are created in the digital world, natural language processing technology has become an important tool that facilitates big data analysis and improves interaction between humans and computers. NLP is an important technology area that performs many tasks, such as text analysis, language translation, text classification, sentiment analysis, and text summarization, using machine learning and AI algorithms.

There are many examples of AI applications in this field. But the first examples that come to mind are definitely text analysis comprehension and translation. AI performs meaning extraction by analyzing large amounts of text data. Natural language processing is used for topics such as sentiment analysis, text classification, meaning extraction, and text summarization. AI eliminates language barriers by automatically translating text into different languages. Machine translation makes it easy to share and understand multilingual content. In addition to all these, studies such as speech and language interfaces, text sequencing and relevance, and fake news and misleading content detection can be listed among the examples that can be given to AI applications in this field.

Natural language processing combined with AI provides many benefits, such as understanding texts, extracting meaning, and interacting with users in a natural language. Applications such as voice assistants, chatbots, and language translation services have entered our lives thanks to AI and natural language processing technologies and play an important role in our daily lives. With the advancement of these technologies, it is expected that even more advanced and effective applications will emerge in the field of natural language processing. However, as in all other AI application areas, it is important to consider the ethical and data privacy issues of these developments and to use artificial intelligence safely and responsibly.

10.3.9　MILITARY, DEFENSE, AND SECURITY AREAS

AI has a wide range of uses and importance in the military and defense fields, as in many areas today. States spend a considerable part of their budgets in the military field every year. Most of the recent investments in this direction are for AI applications. AI has now become one of the priority areas for states in the military and defense field and has critical importance. AI technology has had a major impact in developing and modernizing the security, strategy, and operational processes of military and defense forces. AI has the potential to become a more powerful tool for dealing with security threats by increasing military capabilities in areas such as data analysis, predictability, targeting, and automation. When it comes to AI in military and defense works, the first applications that come to mind include autonomous systems and robot attack-and-defense systems.

In the military field, AI has a wide range of uses for autonomous systems and robots. Unmanned aerial and ground vehicles can perform tasks automatically thanks to AI algorithms. Autonomous robots take on dangerous and challenging tasks, such as reconnaissance, surveillance, and attack, increasing people's safety. AI is used in attack-and-defense systems, such as targeting and guiding weapon systems. The effectiveness of military operations is increased by using AI techniques in processes such as target detection, analyzing security threats, and taking precautions. Enemy and threat analysis can be given as an example of the widespread use of AI in this field. AI provides the ability to understand and analyze enemy movements and threats through big data analytics. The information obtained from the data plays an important role in determining defense strategies and planning operations.

Apart from these, war simulations, combat training, logistics, military resource management, information gathering, intelligence analysis, and other applications can be given as examples of the applications of AI in the fields of military, defense, and security. National defense and security threats are seen as one of the most fundamental problems of all countries. Therefore, the applications of AI in the military and defense fields are of great importance to increase the effectiveness of military forces, reduce security risks, and ensure the safety of personnel. However, paying attention to the ethical and legal issues of these applications is important to responsibly manage the use of AI in the military. The potential of artificial intelligence in the military and defense fields will continue to increase in the coming years, and investments and studies in this field will continue to play an important role in world peace and security.

10.3.10 CYBERSECURITY

Today, the digital world has become an indispensable part of business processes and people's lives. With the rapid development of technology, the security of digital assets and information is becoming increasingly important. Cyberattacks and security threats are becoming more sophisticated day by day, forcing security professionals to take a more proactive and effective approach. At this point, it is thought that the effective use of AI applications in the field of cybersecurity can make the digital world safer and more resilient. The best example of AI applications in cybersecurity is in threat detection and intelligence analysis. AI detects anomalous activities by analyzing network traffic and is used to identify potential threats. AI-based security systems can detect abnormal situations more quickly and precisely by learning normal traffic patterns. In addition, AI has great potential to track cyberattacks and perform intelligence analysis by analyzing large datasets.

Attacks that can be made to another AI-based cybersecurity application can be detected at the first stage and given the ability to take measures quickly and effectively. AI-based security systems are used to detect malware, block malicious connections, and take defensive measures against cyberattacks. Apart from this, AI can use user identities and access rights in verification processes. By analyzing biometric data, AI algorithms can be used more effectively to identify users and protect against attacks such as phishing. It has become imperative for states to protect their assets, develop national and regional cybersecurity policies, train expert personnel, and establish relevant ministries and military units to ensure cybersecurity. Likewise, for companies to compete with their competitors, to protect the confidentiality and privacy of their customers, and to maintain their existence, end user security awareness should be increased, and the level of expertise of cybersecurity personnel should be kept at a high level.

While cybersecurity threats are increasing day by day, the importance of AI technology in the field of cybersecurity is also increasing. AI can work faster and more effectively than humans at detecting and preventing threats. However, the use of AI in the field of cybersecurity also brings some challenges. Issues such as data privacy, ethical issues, and algorithm security are important considerations for the ethical and reliable use of artificial intelligence in cybersecurity. As a result, AI and cybersecurity are representatives of a strong alliance in securing the digital world. With AI, more effective threat detection and defense measures can be taken, while the knowledge and experience of cybersecurity experts contribute to the development of artificial intelligence. This cooperation is of great importance in terms of preventing cyberattacks and making the digital world a safer place.

10.4 ARTIFICIAL INTELLIGENCE AND ETHICS AND MORALS CONCEPTS

Societies can be defined as complex structures where people live together, interact, and share resources. Ethical and moral concepts are of great importance for this social structure to function properly and be sustainable. Basically, the concept of *ethics* is defined as the science that investigates the essence and foundations of morality, and the branch of philosophy that studies the problems related to human behavior. Ethics is based on the fundamental principles that guide people's behavior, values, and beliefs. Acting under these principles helps people fulfill their responsibilities at the individual and societal levels. The benefits of ethical principles for societies in a society are as follows.

- Ethical values ensure that individuals respect each other, are fair, and maintain the social order.
- Ethical values strengthen relations between people.
- Ethical values play an important role in the formation of corporate and social values.
- Ethical values develop individuals' moral and conscience understanding.
- Ethics helps individuals and institutions understand their social responsibilities.
- Ethical principles form the basis of the law and the justice system.
- Societies take steps toward an environmentally and socially sustainable future by adopting ethical values.

The concept of morality, on the other hand, is a system of rules that regulate people's relations with society and with each other. At no time in history have some moral norms and rules had general validity for everyone, and there have been no moral rules that have general validity everywhere and at any time. Therefore, moral rules can change according to conditions and have a flexible structure. Morality is one of the cornerstones of societies and plays a critical role in ensuring the order, harmony, and well-being of society. Moral values strengthen people's relationships with each other, promote social justice, and support the character development of individuals. Societies can create a just and sustainable structure that respects human rights under the guidance of morality. Therefore, the importance of morality in social life is an undeniable fact, and the adoption of moral values by individuals, institutions, and societies forms the basis of social development and welfare.

Today, the concept of ethics is seen as a discipline that examines and regulates behavior patterns in business life. The concept of morality, on the other hand, is seen as a discipline that regulates the relations of people in social life. From this perspective, it can be seen that there is a deep connection between the concepts of ethics and morality. In short, it can be said that ethics is a moral philosophy. Recently, these concepts have started to be discussed in the AI study area, which is being used flawlessly in almost every field of social and business life to ensure the social rules explained in detail earlier and the environmental and social sustainability of societies. In the next section, the problems brought by the field of AI work will be examined, and what kind of concerns they may cause in society will be discussed by giving concrete examples.

10.5 ETHICAL CONCERNS RAISED BY THE CONCEPT OF ARTIFICIAL INTELLIGENCE

The existence, importance, and necessity of ethical values in all areas of life for individuals and societies are an indisputable fact. Especially in this period when social moral collapse is mentioned, it is necessary to review the concept of ethics and keep it up to date in every field. This fact has started to make itself felt in the field of AI, which is effective in all areas of life today. AI is an important field of study that is needed in every field of life, such as social, cultural, economic, and health, and its usage area is becoming widespread day by day. However, lately, it has started to come to the fore with the harm it gives to society and the social order, as well as the benefits it provides in these areas. Recently, large-scale consortia have been established in many countries, and the issue of determining the boundaries of AI has been discussed. Many societies ask, "How far should AI be allowed?" The question began to be spoken loudly, and a large part of society started to think that the ethical and moral problems brought about by AI studies should be tackled. Some of the ethical problems that come with AI are listed in what follow.

10.5.1 Prejudice and Justice

AI applications make decisions based on the data they are trained on. AI systems can also produce biased results if training data is consciously or unconsciously biased. This can lead to discrimination based on gender, ethnicity, or other demographic factors. The diversity and control of the dataset are important for AI to produce fair and unbiased results. For example, when AI algorithms are used in areas such as crime prediction, they may tend to predict future crimes based on past crime data. This can target groups that are already marginalized or prejudiced and lead to further surveillance of these groups. In another example, while AI is used to evaluate candidates in recruitment processes, systems can also be infected with biases, such as gender, ethnicity, or age, resulting from past hiring decisions. This, in turn, can promote inequality and reduce diversity. AI applications are also frequently used in language processing studies. AI-based language processing systems can make predictions about content by analyzing texts. However, such systems can lead to false conclusions by capturing or reinforcing biases in texts.

To give a more concrete example, the ImageNet database has a lot more facial data from people with white skin than with black skin. When we train our AI algorithms to recognize facial features using a database that does not include the correct balance of faces, the algorithm will not work as well on Black faces and will create a built-in bias that can have a large effect. The main reason AI algorithms are biased is that the training data and the algorithms that analyze this data reflect biases. Education data may include general trends in society and existing biases. In addition, unconscious or conscious choices in the design of algorithms or data selection can cause biases to enter AI systems. The potential for AI to be biased is a serious ethical issue. In particular, the tendency of training data and algorithms to reflect biases may limit the wide use of the technology. Therefore, AI developers and researchers should actively seek solutions for bias reduction and control. Measures such as diversity of training data, transparency, and auditing of algorithms can help reduce the potential for AI to be biased and enable the development of more fair, unbiased, and socially acceptable systems.

10.5.2   Responsibility and Accountability

One of the ethical problems brought about by AI studies is undoubtedly responsibility and accountability. The development and use of artificial intelligence usually take place with the contribution of many stakeholders (developers, users, governments, etc.). However, uncertainties about in which situations and who is responsible can complicate the resolution of problems and errors. *Accountability* means that decisions made by AI systems and their results are traceable. The number of examples that can be given to ethical problems in the field of responsibility and accountability of AI applications in the real world is quite high. For example, while working on autonomous driving technology, automobile manufacturers have faced legal and ethical uncertainties about who is responsible for the decision mechanisms of vehicles. An uncertainty that many in this field may have thought of by now is who will be held responsible in such a situation. If an autonomous vehicle is involved in an accident, it can create uncertainty whether the responsibility lies with the manufacturer or the driver.

A striking example of responsibility and accountability in the field of AI can be given in the field of medicine and healthcare. AI-based medical diagnostic systems have the ability to identify diseases and offer treatment recommendations. However, if a disease is misdiagnosed or the treatment recommendation is wrong, it is an important issue who takes responsibility. The responsibility and accountability issues of artificial intelligence are important issues that require us to focus on the ethical and social dimensions of technology. As the examples illustrate, the decisions and consequences of AI systems can have serious consequences. Approaches such as transparency, regulation, and stakeholder collaboration can help overcome the responsibility and accountability issues of AI and allow us to maximize the technology's societal benefits.

10.5.3   Privacy and Data Security

AI studies and applications are powerful tools for analyzing data, learning, and making decisions. Therefore, for AI applications to make healthy choices, the system must be trained with a large amount of data. This data may include users' personal information, habits, and preferences. This raises concerns about data security and privacy. The exposure of data to malicious individuals or its unauthorized use can lead to serious consequences. Recently, the rapid spread and development of this technology have raised concerns about data security and privacy in society. The most concrete examples of privacy and data security in the field of AI work are undoubtedly in the field of medicine and health services. Undoubtedly, health data is an important component of AI applications. However, if patients' medical data is leaked or misused, their privacy may be compromised, and patients' trust may be undermined. In another example, AI assistants working with voice commands can collect important information about users' activities at home or at work. This data can make it vulnerable to privacy breaches and abuse.

Stealing the financial information of people, companies, or institutions can be given as a good example of the abuse of privacy and data security of banking and financial services, which is one of the areas where AI applications are used a lot. AI is used for tasks such as detecting fraud by analyzing financial transactions or making investment recommendations. However, stealing or manipulating financial data can cause great economic damage. Privacy for using data has long been an ethical dilemma of artificial intelligence. If you want to benefit from the blessings that AI brings, it must be trained with certain data. However, the fact that algorithms are constantly collecting data in various ways in their environment to improve themselves also brings security and privacy concerns. As the examples show, data security breaches and privacy issues can have serious consequences. Recommendations such as data encryption, data minimization, transparency, and permission are methods that can be used to address the privacy and data security issues of artificial intelligence. Thanks to these measures, AI technology can gain the trust of society and become available more safely.

10.5.4 AUTOMATION AND LABOR INTERACTION

The rapid development of AI technology is leading to transformations in business and industry. One of these transformations is business automation. While AI can increase productivity by automating business processes, it can also impact the future of the workforce. AI has the ability to automate repetitive and routine tasks. It can perform functions such as data analysis, forecasting, reporting, and even complex decision-making. These automation capabilities can increase the efficiency of companies and leave many people unemployed while allowing people to focus on more strategic and creative tasks. Automation, for example, enables the quick execution of jobs with the use of robots in the manufacturing industry. This can speed up production processes and reduce costs, but it can also reduce the number of workers. In another example, AI-powered chatbots can be used in customer service to answer basic questions and assist customers. This can reduce the workload of human customer service representatives but, in some cases, can result in a lack of human touch.

Finally, AI is utilized in tasks within the financial sector, such as in risk analysis, portfolio management, and fraud detection. This can accelerate business processes while also potentially impacting the roles of certain financial analysts. AI is transforming the business world alongside process automation. In the near future, it is widely accepted that AI will automate many jobs, replacing human roles in the workforce and causing significant changes in the business landscape. To facilitate a respectful and honorable transition, there needs to be sufficient societal awareness and education about AI, promoting collaboration between intelligent machines and humans. As demonstrated in the examples, automation can enhance efficiency while simultaneously affecting the labor force. Approaches such as AI training, human–AI collaboration, and social support are vital for managing workforce interactions and preparing for the future. Through these means, the integration of AI technology into the business world and society can be optimally achieved.

10.5.5 DECISION-MAKING AND TRANSPARENCY

AI systems can make complex decisions by analyzing large amounts of data. However, these decisions are often based on complex algorithm structures and can be difficult for humans to understand. The complex algorithms and learning methods of AI systems can make decision-making processes difficult to understand and monitor. While AI offers great potential in many areas, decision-making and transparency issues raise serious ethical concerns. This causes us to question the transparency of the decision-making processes of artificial intelligence. For example, when evaluating loan applications using AI, it is often not transparent what factors algorithms rely on and how they make decisions. Applicants may have difficulty understanding why a loan was denied or approved. In another example, AI algorithms can be employed to support sentencing and parole decisions in courts. However, it is often unclear how these decisions are made and what factors they are based on. This undermines the transparency of justice.

Lastly, AI can be utilized in medical diagnoses to detect diseases. However, when the basis for a diagnosis or the reasons behind a particular decision are not disclosed, its reliability can be questioned by patients and healthcare professionals alike. Issues of decision-making and transparency in AI applications underscore important ethical dimensions of the technology. As seen in the examples, the complexity of decision-making processes and the lack of transparency undermine reliability and fairness. Establishing explainability, traceability, and ethical standards can contribute to making artificial intelligence decision-making processes more understandable and transparent. Thus, the creation of a more ethical and socially acceptable utilization of AI technology becomes possible.

10.5.6 BALANCE OF POWER AND UNFAIR COMPETITION

AI, while causing significant changes in the business world and society at large, must grapple with the ethical issues it brings along. One of these ethical issues is undoubtedly the balance of power and unfair competition among companies and institutions. To briefly explain this, AI algorithms and applications can make decisions based on large amounts of data through data analysis and learning capabilities. This can lead to certain companies or individuals having more access to resources or data compared to others. This situation can disrupt the balance of power and result in unfair advantages. Consequently, in any free market, it may lead to the undesirable concentration of power under a single entity and the bankruptcy of other competing firms. For instance, major technology companies can optimize marketing strategies with AI-based advertising. This could enable companies with larger budgets to create more effective and targeted advertisements, potentially reducing the competitive advantage of small businesses. In another example, AI algorithms are used for high-speed trading in financial markets. Large financial institutions can utilize these algorithms to execute faster and more efficient trades, while small investors might struggle to compete at the same speed.

Lastly, in another example, AI algorithms can be used in recruitment processes to evaluate candidates. However, the fact that large companies have more data and use AI applications efficiently may cause small businesses to be unable to compete with these companies at the same level in their recruitment processes. The issues of power balance and unfair competition arising from AI algorithms underscore a significant ethical concern that highlights the societal and economic impacts of the technology. As evident from the examples, the potential for certain actors to possess greater power or resources carries the risk of impairing competition and innovation. Measures such as regulatory oversight, data equality, and the establishment of ethical standards are approaches that can be employed to address the issues of power balance and unfair competition brought about by artificial intelligence. In this way, it might become possible to realize the societal benefits of AI technology more equitably and sustainably.

The rapid advancement of AI technology has brought forth a multitude of ethical issues. When considering AI, bias, justice and accountability, privacy, data security, automation and human workforce interaction, decision-making, transparency, disruption of power balance, and responsibility are among the ethical concerns that come to mind and require prompt resolution. However, when it comes to the ethical dilemmas brought about by AI, ethical challenges undoubtedly extend beyond these aforementioned areas. Discovered through a Delphi study, Stahl compiled all ethical issues related to AI into a total of 39 categories in his work titled "Ethical Issues of AI" [25]: (1) cost to innovation, (2) harm to physical integrity, (3) lack of access to public services, (4) lack of trust, (5) "awakening" of AI, (6) security problems, (7) lack of quality data, (8) disappearance of jobs, (9) power asymmetries, (10) negative impact on health, (11) problems of integrity, (12) lack of accuracy of data, (13) lack of privacy, (14) lack of transparency, (15) potential for military use, (16) lack of informed consent, (17) bias and discrimination, (18) unfairness, (19) unequal power relations, (20) misuse of personal data, (21) negative impact on the justice system, (22) negative impact on democracy, (23) potential for criminal and malicious use, (24) loss of freedom and individual autonomy, (25) contested ownership of data, (26) reduction of human contact, (27) problems of control and use of data and systems, (28) lack of accuracy of predictive recommendations, (29) lack of accuracy of non-individual recommendations, (30) concentration of economic power, (31) violation of fundamental human rights in the supply chain, (32) violation of fundamental human rights of end users, (33) unintended, unforeseeable adverse impacts, (34) prioritization of the "wrong" problems, (35) negative impact on vulnerable groups, (36) lack of accountability and liability, (37) negative impact on the environment, (38) loss of human decision-making, and (39) lack of access to and freedom of information.

Considering all these ethical issues, the design, implementation, and management of AI systems involve deeper and more crucial considerations. In the future, the development of ethical standards, updating existing legal frameworks, and taking steps toward resolving ethical dilemmas in the field of AI should involve broad societal participation. This way, AI technology can be harnessed for the benefit of society while minimizing adverse impacts.

10.6 PREDICTIVE JUSTICE WITH AI CONCEPTS, AND COPE WITH THEIR PROBLEMS

In general, more conservative societies, or companies that suffer from unfair competition from AI applications in the business world, agree not to use AI applications or to impose strict restrictions, which are needed in many areas due to ethical concerns caused by AI algorithms. However, this strict attitude toward AI applications also means not taking advantage of many of the advantages that AI offers to human beings. At this point, it is necessary to compare how much benefit and how much ethical limits are violated for each application. In order to make this comparison, it would be appropriate to determine the ethical standards and norms within the field of AI work by experts. For these standards and norms to be determined, first of all, the definition of *ethical AI* must be clearly defined without any ambiguity. Before discussing all this, "What is ethics AI?" and how it should be, this question must be answered.

With its current definition, *ethical artificial intelligence* (ethics AI) is an approach that emphasizes ethical values, principles, and standards in the development, use, and application of these technologies. Ethics AI aims to increase the positive effects of technology on society, individuals, and the environment and to minimize the negative consequences. This approach addresses AI ethical issues and provides a framework for the development of technology in a societal, cultural, and people-oriented manner. At this point, one of the most important questions to be asked is who should determine this framework. To cope with the ethical problems that come with artificial intelligence and to search for solutions, all stakeholders must act together. In this consortium to be established among the stakeholders, it is vital for the sustainability of the outputs to be obtained and the solutions to be found that there are representatives from many sectors besides field experts, lawyers, politicians, and legislators.

To protect social and ethical values while benefiting from the advantages of AI, ethical problems should be resolved as much as possible. To solve the ethical problems of AI, consortia are established and discussed in many different parts of the world. Some of the solutions that emerged as a result of these discussions and may be subject to AI standards in the future are listed as follows.

1. *Transparency of AI algorithm decision-making processes.* The decision-making processes of AI algorithms should be made understandable. The basis of algorithms, how they learn from data, and why they make specific decisions can be explained in a clear manner.
2. *Ethical standards developed with the participation of AI stakeholders.* Ethical standards created with the participation of all stakeholders relevant to the field of AI can allow its use in all areas without undermining societal ethics and without causing harm to personal privacy.

3. *Ensuring equality in data access.* Ensuring equal access to data can enhance the competitive advantage of smaller businesses and individuals. This can lead to a level playing field, where everyone has the same opportunities.

4. *Continuous education requirement for adapting to new technologies.* People may need to continually receive education to adapt to new technologies and keep up with evolving requirements. Ongoing education is necessary for individuals to adapt to new technologies and changing demands.

5. *Encryption of data used by AI algorithms.* Encrypting the data used by AI algorithms is an effective method to enhance data security. Encrypted data can be protected from unauthorized access and stored securely.

6. *Transparency in the design and operation of AI algorithms.* Transparency is important in the design and operation of AI algorithms. Having traceable decision mechanisms and algorithm outcomes can aid in error detection and in determining responsibility.

7. *Minimization of unnecessary data collection or storage.* Not collecting or storing unnecessary data can reduce privacy risks. Using only essential data can be a more secure approach to data security.

8. *Transparent user information.* Users should be informed transparently about how their data is used and processed. User consent can make the processes of data collection and usage more transparent.

9. *Collaboration of multiple stakeholders.* Developing and using AI technology require collaboration among various stakeholders. Developers, users, governments, NGOs, and other parties can address responsibilities and accountability more effectively with active participation.

10. *Diverse training data to prevent bias.* During the training of AI algorithms, diversifying training data as much as possible can prevent biased decisions from being introduced into the system.

It is now a reality accepted by everyone that some standards need to be established to cope with the ethical problems caused by AI algorithms. Worldwide, the list given can be considered as the beginning of the framework and norms that will be established for the development of algorithms and applications for the development of AI-based fair, neutral, and socially acceptable systems. These established standards can be updated as needed and as the AI perspective changes in the future.

10.7 CONCLUSIONS AND THE FUTURE OF PREDICTIVE JUSTICE WITH AI

In its simplest form, *AI* is a field of science and technology that attempts to simulate the human-like thinking and decision-making abilities of computer systems. The main purpose of AI algorithms is to enable computers to perform human-like abilities, such as learning, inference, problem-solving, language comprehension, image processing, voice recognition, and decision-making. AI is a technology that is becoming more and more widespread in all areas of life today. As the field of AI work increases, the dependence of business life on it increases as well, making it an indispensable condition. However, there are ethical problems caused by AI applications as well as the benefits they provide. These problems have recently begun to be voiced loudly, especially in the business world. Although especially private companies and institutions that carry out commercial activities are eager to use the innovations and advantages of artificial intelligence, they are also concerned about the potential risks that artificial intelligence may cause in the future. The source of these concerns is the possibility of situations with serious legal dimensions, such as the deprivation of rights, confidentiality, and security, which may endanger their commercial titles and cause great financial and reputational losses. In this regard, the main purpose is to ensure the development of socially acceptable systems while making maximum use of AI algorithms.

In this study, the possible risks that AI algorithms and applications that we see frequently in all areas of life may cause in the near future have been discussed, and some solutions have been presented. It has been accepted by the public that some standards and norms should be established to cope with the ethical problems caused by AI. The fact that these standards, which are planned to be created, must be generally accepted is of vital importance in terms of their applicability. Establishing a consortium consisting of different segments of society, different sectors of working life, and different nationalities to establish these standards will ensure general acceptability. Creating regulations and standards that control the use of AI can enable legislators to take the necessary measures for these problems that are contrary to the moral values of society and may disturb peace.

Apart from this, the ethical problems brought by the field of AI work today have become a global problem rather than a problem of a particular state, race, or region. Therefore, the presence of experts from different regions and countries in the consortium to be formed ensures that the definition and limitations of ethics AI can be more general and acceptable. Besides, every nation has its own values and moral limits. Therefore, the presence of stakeholders from different nationalities in the consortium to be formed will increase the inclusiveness and acceptability of ethical AI worldwide. The ethical problems of artificial intelligence may become more complex as technology advances. The field of AI is developing very rapidly due to the nature of technology. These standards need to be updated and improved as AI algorithms evolve and their application areas expand. In the future, steps should be taken to solve ethical problems in AI with the development of ethical standards, the update of existing legal frameworks, and the broad participation of society. Therefore, more work, effort, and collaboration will be required to tackle AI ethical issues in the future.

REFERENCES

1. Turing, A. M. (1950). Computing Machinery and Intelligence. *Mind*, 59(236), 433–460. doi: 10.1093/mind/LIX.236.433.
2. McCarthy, J. (2007). From Here to Human-Level AI. *Artificial Intelligence*, 171(18), 1174–1182. doi: 10.1016/j.artint.2007.10.009.
3. Russell, S. J., & Norvig, P. (2016). *Artificial Intelligence: A Modern Approach* (3rd ed.). ABD, Pearson. ISBN: 978-0-13-604259-4.
4. Ergen, M. (2019). What is Artificial Intelligence? Technical Considerations and Future Perception. *The Anatolian Journal of Cardiology*, 22(2), 5–7. doi: 10.14744/AnatolJCardiol.2019.79091.

5. Duncan, G. M., & Couturat, L. (1903). La Logique de Leibniz d'apres des Documents Inedits. *The Philosophical Review*, 12(6), 649–664. doi: 10.2307/2176982.
6. Church, A. (1936). An Unsolvable Problem of Elementary Number Theory. *American Journal of Mathematics*, 58(2), 345–363. doi: 10.2307/2371045.
7. Church, A. (1936). A Note on the Entscheidungsproblem. *Journal of Symbolic Logic*, 1(1), 40–41. doi: 10.2307/2269326.
8. Turing, A. M. (1937). On Computable Numbers, with an Application to the Entscheidungs Problem. *Proceedings of the London Mathematical Society*, 42(1), 230–265. doi: 10.1112/plms/s2-42.1.230.
9. Turing, A. (1951). Intelligent Machinery: A Heretical Theory. In J. Copeland (ed.) *The Essential Turing*, 472–475. Oxford, Great Britain: Clarendon Press.
10. Muggleton, S. (2014). Alan Turing and the Development of Artificial Intelligence. *AI Communications*, 27(1), 3–10. doi: 10.3233/AIC-130579.
11. Samuel, A. L. (1959). Some Studies in Machine Learning Using the Game of Checkers. *IBM Journal of Research and Development*, 3(3), 210–229. doi: 10.1147/rd.33.0210.
12. Krishna, C. V., Rohit, H. R., & Mohana. (2018). A Review of Artificial Intelligence Methods for Data Science and Data Analytics: Applications and Research Challenges. In *2nd International Conference on I-SMAC (IoT in Social, Mobile, Analytics and Cloud)*, 591–594. IEEE. doi: 10.1109/I-SMAC.2018.8653670.
13. Ulhaq, A. (2021). Deep Learning, Past Present and Future: An Odyssey. *Machine Vision and Digital Health Research Group*, 1(1), 1–7. doi: 10.31224/osf.io/vrmk4.
14. Hinton, G. E. (1980). Inferring the Meaning of Direct Perception. *Behavioral and Brain Sciences*, 3(3), 387–388. doi: 10.1017/S0140525X00005549.
15. LeCun, Y., Boser, B., Denker, J. S., Henderson, D., Howard, R. E., Hubbard, W., & Jackel, L. D. (1989). Backpropagation Applied to Handwritten Zip Code Recognition. *Neural Computation*, 1(4), 541–551. doi: 10.1162/neco.1989.1.4.541.
16. Boser, B. E., Guyon, I. M., & Vapnik, V. N. (1992). A training algorithm for optimal margin classifiers. In *Proceedings of the Fifth Annual Workshop on Computational Learning Theory*, 144–152. ACM. doi: 10.1145/130385.130401.
17. Hochreiter, S., & Schmidhuber, J. (1997). Long Short-Term Memory. *Neural Computation*, 9(8), 1735–1780. doi: 10.1162/neco.1997.9.8.1735.
18. Hinton, G. E., Osindero, S., & Teh, Y.-W. (2006). A Fast-Learning Algorithm for Deep Belief Nets. *Neural Computation*, 18(7), 1527–1554. doi: 10.1162/neco.2006.18.7.1527.
19. Krizhevsky, A., Sutskever, I., & Hinton, G. E. (2012). Imagenet Classification with Deep Convolutional Neural Networks. In *Advances in Neural Information Processing Systems*, 25–34. ABD.
20. Goodfellow, I. J. (2014). *On Distinguishability Criteria for Estimating Generative Models*. http://arxiv.org/abs/1412.6515 [accessed May 2, 2011].
21. He, K., Zhang, X., Ren, S., & Sun, J. (2016). Identity mappings in deep residual networks. In *Proceedings of the Computer Vision–ECCV 2016: 14th European Conference*, 630–645. Springer International Publishing.
22. Yu, H. (2016). From Deep Blue to DeepMind: What AlphaGo Tells Us. *Predictive Analytics and Futurism*, 13(1), 42–45.
23. Mikolov, T., Grave, E., Bojanowski, P., Puhrsch, C., & Joulin, A. (2017). *Advances in Pre-Training Distributed Word Representations*. arXiv Preprint, arXiv:1712.09405. doi: 10.48550/arXiv.1712.09405.
24. Jumper, J., Evans, R., Pritzel, A., Green, T., Figurnov, M., Tunyasuvunakool, K., & Hassabis, D. (2020). *AlphaFold 2. Fourteenth Critical Assessment of Techniques for Protein Structure Prediction*. DeepMind.
25. Stahl, B. C. (2021). Ethical Issues of AI. In *Artificial Intelligence for a Better Future. Springer Briefs in Research and Innovation Governance*, 35–53. Springer. doi: 10.1007/978-3-030-69978-9_4.

11 Integrating Cybersecurity in the Design and Implementation of Intelligent and Sustainable Manufacturing Systems

M. Baritha Begum

11.1 INTRODUCTION

The integration of smart technologies and sustainable practices into manufacturing processes has given rise to intelligent and sustainable manufacturing. *Sustainable manufacturing* attempts to use less resources, support renewable energy sources, and lower greenhouse gas emissions, whereas *intelligent manufacturing* makes use of data analytics, AI, robots, and automation to increase efficiency. However, with the adoption of these technologies comes the need for robust cybersecurity measures to mitigate the associated risks [1].

Risks and Challenges. One of the main challenges in cybersecurity for intelligent and sustainable manufacturing is the interconnected nature of the systems used in manufacturing processes. IoT devices, cloud computing, and AI systems rely on interconnected systems to collect and analyze data in real time. Cybercriminals may use these newly created vulnerabilities to disrupt business processes or obtain unauthorized access to critical data. Additionally, it may be difficult to build a thorough cybersecurity strategy due to the absence of standardization in security methods and practices [2, 3].

Another challenge is the shortage of skilled cybersecurity professionals. With the growing demand for cybersecurity professionals across different sectors, there is a shortage of skilled labor, which can make it difficult for businesses to secure their systems effectively [4].

Best Practices. To mitigate the cybersecurity risks associated with intelligent and sustainable manufacturing, businesses should adopt the following best practices:

DOI: 10.1201/9781003501152-11

(1) Conduct a risk assessment. Create a complete cybersecurity strategy by doing a risk assessment. Businesses should do this to find any potential vulnerabilities [5]. (2) Implement strong authentication protocols. Multi-factor authentication, for example, can assist in preventing unauthorized access to sensitive data [6]. (3) Encrypt sensitive data. By preventing unauthorized access and ensuring secrecy, encryption can assist in safeguarding sensitive data [7, 8]. (4) Train employees. Regular training for employees can help raise awareness about cybersecurity risks and ensure that they follow best practices for maintaining cybersecurity. (5) Partner up. Companies should collaborate closely with their partners, including suppliers and customers, to make sure that everyone is aware of the dangers and has the necessary cybersecurity protections in place [9].

Highlights of the Proposed System

1. Cybersecurity is crucial in intelligent and sustainable manufacturing systems.
2. The integration of cyber-physical systems (CPS) in manufacturing systems creates vulnerabilities that can be exploited by cyberattackers.
3. Effective cybersecurity strategies are necessary to protect manufacturing systems from cyber threats.
4. The application of machine learning (ML) and artificial intelligence (AI) in cybersecurity can improve the identification of and reaction to online threats.
5. Collaboration among stakeholders, including manufacturers, government agencies, and cybersecurity experts, is essential for effective cybersecurity in manufacturing systems.

11.1.1 Definition of Intelligent and Sustainable Manufacturing

Innovative technologies are incorporated into the manufacturing process to produce a more efficient, flexible, and environmentally friendly production system, which is referred to as *intelligent and sustainable manufacturing*. Big data, artificial intelligence, and the Internet of Things are a few examples of these technologies. Optimizing resource utilization, reducing waste, and lessening the negative effects of manufacturing activities on the environment are the main objectives of intelligent and sustainable manufacturing. This is achieved through real-time monitoring and the control of production processes, the use of predictive analytics to optimize decision-making, and the implementation of sustainable practices, such as energy-efficient production and the use of renewable energy sources. Ultimately, intelligent and sustainable manufacturing represents a step forward in the evolution of the manufacturing sector, one that leverages technology to create a more sustainable and efficient production system [10].

11.1.2 The Importance of Cybersecurity in Intelligent and Sustainable Manufacturing

It is impossible to exaggerate the value of cybersecurity in intelligent and environmentally friendly production. The number of possible security threats and vulnerabilities has expanded along with the rising integration of cutting-edge technologies into the manufacturing process, including IoT, AI, and big data. Cyberattacks can result in significant harm to manufacturing operations, including loss of sensitive data, disruptions to production processes, and financial losses. Additionally, the use of IoT and other connected devices in the manufacturing process creates an increased risk of unauthorized access to systems and data, making cybersecurity a critical concern for manufacturers [11]. Cybersecurity is also essential for ensuring the stability and reliability of intelligent and sustainable manufacturing systems. With the increasing use of AI and big data in the production process, it is important to ensure that these systems are secure from potential malicious actors who may try to manipulate or exploit them for their own gain. By implementing robust cybersecurity measures, manufacturers can ensure the integrity and accuracy of their systems and minimize the risk of disruptive incidents that could have serious consequences for their operations [12].

Moreover, cybersecurity is also important for maintaining trust in the manufacturing industry. As consumers become increasingly aware of the importance of protecting their data, they are more likely to choose products from companies that prioritize the security of their systems and data. By investing in cybersecurity, manufacturers can demonstrate their commitment to the responsible use of technology and help build trust with customers, partners, and other stakeholders.

11.1.3 THE IMPORTANCE OF CYBERSECURITY IN INTELLIGENT AND SUSTAINABLE MANUFACTURING SYSTEMS

Cybersecurity is crucial in intelligent and sustainable manufacturing systems because of the increasing integration of CPS, which creates new vulnerabilities for cyberattackers to exploit. Cyberattacks on manufacturing systems can lead to significant financial losses, operational disruptions, and compromise of sensitive data. Moreover, cyberattacks on critical infrastructure, including manufacturing systems, can have serious consequences for public safety and national security. Therefore, effective cybersecurity strategies must be implemented to protect manufacturing systems from cyber threats [13].

11.1.4 FUTURE DIRECTIONS AND CHALLENGES

In the future, cybersecurity in manufacturing systems will continue to be a critical issue as technology advances and cyber threats become more sophisticated. Some of the challenges facing cybersecurity in manufacturing systems include the lack of cybersecurity expertise among manufacturers, the need for increased collaboration and information sharing among stakeholders, and the evolving nature of cyber threats. Moreover, the use of AI and ML in cybersecurity presents both opportunities and challenges, as these technologies can enhance the detection and response to cyber threats but also introduce new vulnerabilities. To address these challenges, ongoing research, training, and collaboration among stakeholders are necessary to ensure the cybersecurity of intelligent and sustainable manufacturing systems.

11.2   UNDERSTANDING THE THREAT LANDSCAPE

11.2.1   Types of Cybersecurity Threats Faced by Manufacturing Systems

Manufacturing systems face a wide range of cybersecurity threats, including:

1. *Malware attacks.* One type of destructive software designed to harm or steal data from computer systems is malware, which is used in attacks. This can contain Trojans, worms, viruses, and ransomware.
2. *Phishing attacks.* In order to trick the recipient into disclosing personal information, such as usernames, passwords, and credit card details, phony emails that appear to be from a trustworthy source are sent.
3. *Denial-of-service assaults.* Denial-of-service assaults entail overwhelming a system with traffic or requests, which can lead to the system crashing or going offline.
4. *Insider threats.* An insider threat is the potential for users who have been granted access to a system, such as employees, contractors, or other users, to maliciously or accidentally damage the system or its data.
5. *Advanced persistent threats.* Sophisticated attacks that are deliberately directed at a single company or person are known as advanced persistent threats (APTs). They can combine several strategies, such as malware, social engineering, and others.

11.2.2   Identification of Critical Assets and Potential Vulnerabilities

To effectively manage cybersecurity risks, it is important to identify critical assets and potential vulnerabilities. This can involve conducting a thorough inventory of all hardware and software systems, as well as any data that is stored or processed by these systems. Some key steps to identify critical assets and potential vulnerabilities include:

1. *Conducting a risk assessment.* This involves identifying the potential threats and vulnerabilities that could affect the manufacturing system, as well as the potential impact of these risks on the organization.
2. *Identifying critical assets.* Critical assets may include data, systems, equipment, and other resources that are essential to the functioning of the manufacturing system. These assets should be prioritized for protection.
3. *Assessing vulnerabilities.* Vulnerabilities can include outdated software, weak passwords, unsecured wireless networks, and other factors that can be exploited by attackers.

11.2.3   Vulnerabilities in Intelligent and Sustainable Manufacturing

- *Internet of Things (IoT) devices.* IoT devices are often deployed without proper security measures and can be vulnerable to attacks that exploit weak passwords, unpatched software, or other vulnerabilities. Attackers can compromise IoT devices to gain access to the network or disrupt operations.

- *Cloud computing.* The use of cloud computing in manufacturing can introduce new security risks, such as data breaches, unauthorized access, and denial-of-service attacks. Cloud providers can also be a target for attacks, which can impact the availability of cloud services.
- *Supply chain security.* Manufacturing supply chains are often complex and involve multiple parties, making them vulnerable to attacks, such as supply chain attacks, where attackers compromise a vendor to gain access to the manufacturer's network.
- *Legacy systems.* Legacy systems are often outdated and no longer supported, making them vulnerable to attacks that exploit known vulnerabilities. Upgrading these systems can be costly and time-consuming, leading some organizations to continue using them.

Employees may unwittingly add security risks through their activities, such as clicking on phishing emails or using weak passwords, due to a lack of awareness and training. Organizations may become exposed to assaults if they lack cybersecurity education and training.

In order to protect their industrial activities, organizations should take the necessary cybersecurity precautions and be aware of these threats and vulnerabilities.

11.2.4 BEST PRACTICES FOR CYBERSECURITY IN INTELLIGENT AND SUSTAINABLE MANUFACTURING

Once critical assets and potential vulnerabilities have been identified, it is important to conduct a comprehensive risk assessment and management process. This can involve:

1. Evaluating the likelihood and potential impact of each identified risk.
2. Developing a risk management plan that outlines strategies for mitigating each risk.
3. Implementing appropriate security controls and measures to reduce the likelihood of a successful attack.
4. Monitoring and reviewing the effectiveness of security measures over time, and making adjustments as needed.

Overall, effective risk assessment and management are critical components of a comprehensive cybersecurity strategy for manufacturing systems. By identifying critical assets and potential vulnerabilities and implementing appropriate security measures, organizations can help protect their manufacturing systems from a wide range of cybersecurity threats.

11.2.4.1 Risk Assessment and Management

Conducting a risk assessment is an essential step in developing a cybersecurity strategy. Organizations should identify their assets, the risks they face, and the potential impact of a breach. Risk management involves prioritizing risks and implementing appropriate controls to mitigate them. This can include regular vulnerability scanning and penetration testing, implementing access controls and firewalls, and maintaining up-to-date software and hardware.

11.2.4.2 Network Segmentation and Access Control

Segmentation of the network can help limit the impact of a cybersecurity incident by restricting the spread of malware and other threats. Organizations can segment their networks based on the sensitivity of their data and implement access controls and firewalls to prevent unauthorized access. This can help reduce the risk of lateral movement of threats across the network.

11.2.4.3 Encryption and Authentication

For the protection of sensitive data, encryption and authentication are necessary. Data can be protected while it is in motion or at rest using encryption, and unauthorized access can be avoided with the aid of authentication. Organizations should implement strong encryption protocols, such as SSL/TLS and VPNs, and use multi-factor authentication to improve access controls.

11.2.4.4 Incident Response Plan

An incident response strategy explains what should be done in the event of a cybersecurity issue. Processes for recognizing and reporting incidents, containing the problem, and recovering from the incident should all be part of this. The plan should also identify roles and responsibilities, including incident response team members and management.

11.2.4.5 Security Awareness and Training

The weakest link in cybersecurity is frequently employees. Employee education about the value of cybersecurity and how to spot and respond to possible attacks can be aided by ongoing training and awareness efforts. This can include training on password management, identifying phishing emails, and reporting suspicious activity.

Overall, the aforementioned best practices should be implemented in conjunction with each other to ensure a robust cybersecurity strategy. By prioritizing cybersecurity, organizations can reduce the risk of cyberattacks and ensure the security and sustainability of their manufacturing operations.

11.3 CYBERSECURITY REQUIREMENTS IN THE DESIGN OF INTELLIGENT AND SUSTAINABLE MANUFACTURING SYSTEMS

11.3.1 Secure Architecture Design Principles

In the context of intelligent and sustainable manufacturing systems, secure architecture design principles involve designing systems that are resilient to cyberattacks and have multiple layers of protection. This requires a comprehensive understanding of the system's functional requirements, data flow, and information security needs. Some key principles of secure architecture design include the following.

11.3.1.1 Defense in Depth

To defend against various assaults, this means implementing many layers of security measures. For instance, access control measures can be used to restrict unauthorized access to the system, while encryption and data masking can be used to protect sensitive data.

Defense in depth is a security principle that calls for adding additional security measures to a system to shield it from potential attackers. It becomes increasingly challenging for an attacker to breach the system as each layer adds a layer of defense.

A suitable equation that represents defense in depth can be:

security = (security control 1 × security control 2 × security control 3 × ... × security control n) (1)

In this equation (1), each security control represents a layer of defense, and the product of all the security controls represents the overall security of the system. The more security controls there are, the more secure the system is likely to be.

Defense in depth manufacturing systems, for instance, may incorporate many levels of security controls, including firewalls, intrusion detection systems, access controls, encryption, and routine security audits. By using multiple layers of security controls, the system is better protected against potential threats, even if one layer of defense is breached. A defense in depth diagram typically consists of multiple layers of security controls that are implemented to protect a system. Each layer offers a distinct kind of defense, and together they form an all-encompassing security plan. Here is a defense in depth diagram as an illustration:

In Figure 11.1, the first layer is *physical security*, which includes measures such as surveillance cameras, guards, and access controls to physically secure the manufacturing system's facilities. The second layer is *network perimeter*, which includes firewalls and intrusion detection systems to protect against unauthorized access from external sources. The third layer is *access control and authentication*, which includes measures such as passwords and biometrics to control access to the system. The fourth layer is *data encryption*, which includes technologies such as SSL, IPSEC, and VPNs to protect data in transit. The final layer is *security information and event management*, which includes log analysis and reporting to monitor the system for security incidents and generate alerts when necessary.

By implementing multiple layers of security controls like these, defense in depth can help protect manufacturing systems from a wide range of potential threats and vulnerabilities.

11.3.1.2 Separation of Duties

In order to avoid any one person from having an excessive amount of control over the system, this idea calls for dividing up the duties and responsibilities among several people or organizations. This can help reduce the risk of insider threats and limit the impact of any security breaches. For example, in a manufacturing system, separation of duties might involve dividing responsibilities among different teams or individuals. One team might be responsible for designing and testing the system, while another team might be responsible for implementing and maintaining it. Additionally, access

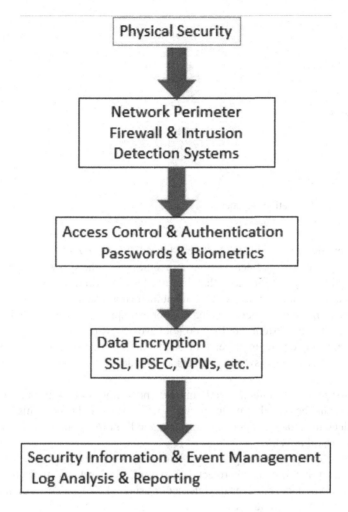

FIGURE 11.1   Block diagram of defense in depth.

controls might be implemented to restrict access to certain functions of the system to specific users, preventing any one person from having too much control.

The principle of separation of duties is often implemented in conjunction with other security measures, such as access controls, authentication, and auditing. Together, these measures help ensure that the system is secure and protected against potential threats, while also promoting accountability and transparency in the system's operations.

Here is an example of a block diagram (Figure 11.2) that illustrates the principle of separation of duties in a manufacturing system. The manufacturing system is divided into three main teams: *design*, *testing*, and *operations*. Each team is responsible for a different aspect of the system and has different levels of access to the system's functions.

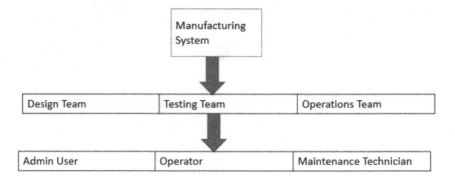

FIGURE 11.2 Separation of duties.

Additionally, there are three different types of users: *admin user, operator,* and *maintenance technician.* Each user has a specific role in the system, and their access is restricted to only the functions that they need to perform their duties.

By separating duties in this way, the manufacturing system is more secure and less vulnerable to fraud or malicious behavior. For example, the design team is responsible for creating the system, but they do not have access to the operations functions. This prevents any one team or person from having too much control over the system and helps ensure that the system operates as intended.

1. *Least privilege.* This principle involves providing users with only the permissions they need to perform their specific roles and tasks. This can help reduce the attack surface of the system and limit the damage caused by any successful attacks.
2. *Secure communication protocols.* This principle involves using secure communication protocols to protect data in transit. For instance, using SSL/TLS to encrypt data sent over the internet or using VPNs to secure communication between remote sites.

11.3.2 Security by Design (SBD) Approach

An alternative to developing systems with security as an afterthought is to use the security by design (SBD) method. This involves incorporating security requirements into the system design process and ensuring that security is an integral part of the system's overall functionality. Some key principles of the SBD approach include:

1. *Threat modeling.* This involves identifying potential threats to the system and designing security controls to mitigate those threats. Threat modeling can help identify vulnerabilities and design systems to prevent or reduce the impact of attacks.
2. *Secure coding practices.* This involves writing code that is secure by default and incorporating security best practices into the coding process. This can help prevent common coding errors that can be exploited by attackers.

3. *Regular security testing.* This involves testing the system for vulnerabilities on a regular basis and ensuring that any issues are addressed promptly. Regular security testing can help identify new vulnerabilities and ensure that the system remains secure over time.

11.3.3 Integration of Cybersecurity in the System Development Life Cycle (SDLC)

Integrating cybersecurity into the system development life cycle (SDLC) involves ensuring that security is considered at every stage of the development process. This includes planning, design, implementation, testing, and maintenance. Some key principles of integrating cybersecurity into the SDLC include:

1. *Establishing clear security requirements.* This involves defining the security requirements for the system at the outset of the project. Clear security requirements can help ensure that security is a priority throughout the development process.
2. *Incorporating security into the design process.* This involves designing the system with security in mind from the outset, rather than as an afterthought. This can help ensure that security is an integral part of the system's overall functionality.
3. *Testing for security vulnerabilities.* This involves testing the system for security vulnerabilities on a regular basis and ensuring that any issues are addressed promptly. This can help ensure that the system remains secure over time.

11.3.4 Compliance with Cybersecurity Standards and Regulations

It involves ensuring that the system meets the requirements of relevant security standards and regulations. This includes industry-specific standards, such as ISO 27001, as well as more general standards, such as NIST and CIS. Some key principles of compliance with cybersecurity standards and regulations include:

- *Understanding the relevant standards and regulations.* This involves understanding the requirements of relevant security standards and regulations and ensuring that the system meets those requirements.
- *Conducting regular audits.* This involves conducting regular audits of the system.
- *National Institute of Standards and Technology (NIST) Cybersecurity Framework.* The NIST Cybersecurity Framework is a set of guidelines for organizations to manage and reduce cybersecurity risks. The framework is voluntary but widely adopted in the United States and internationally. It consists of five core functions—identify, protect, detect, respond, and recover—and provides guidance on implementing best practices for each function.
- *International Organization for Standardization (ISO) 27001 and 27002.* These are international standards that provide a framework for establishing, implementing, maintaining, and continually improving an information security management system. ISO 27001 provides the requirements for

an information security management system, whereas ISO 27002 offers instructions for putting the system into place and keeping it up-to-date.

- *The European Union Institution for Cybersecurity (ENISA).* ENISA is an EU institution in charge of advancing cybersecurity within the EU. ENISA provides guidance and recommendations for improving cybersecurity in various sectors, including manufacturing.
- *Cybersecurity Information Exchange Act (CISA).* Enacted in the United States in 2015, CISA is a law that promotes the exchange of cybersecurity data between the public and private sectors. It provides liability protections for private entities that share information with the Department of Homeland Security and other government agencies.
- *General Data Protection Regulation (GDPR).* In the European Union, the collection, use, and protection of personal data are governed by the General Data Protection Regulation (GDPR). Any entity that handles the personal data of EU citizens is required to abide by the rule, regardless of where it is situated. Regulations for data security and protection are included in the GDPR, such as the requirement to notify individuals of data breaches within 72 hours.

Overall, adherence to these regulations and standards can help organizations establish a strong cybersecurity framework and reduce the risk of cyberattacks in intelligent and sustainable manufacturing. By following these guidelines, organizations can improve their cybersecurity posture and ensure the protection of sensitive data and systems.

11.4   CYBERSECURITY CONTROLS IN THE IMPLEMENTATION OF INTELLIGENT AND SUSTAINABLE MANUFACTURING SYSTEMS

11.4.1   Access Control and Identity Management

Access control and identity management are critical cybersecurity controls for intelligent and sustainable manufacturing systems. *Access control* ensures that only authorized personnel have access to the system, while *identity management* ensures that the system can verify the identity of the authorized personnel. These controls help prevent unauthorized access and ensure that the system is secure.

11.4.2   Network Security and Segmentation

Network security and segmentation are important cybersecurity controls that help protect intelligent and sustainable manufacturing systems from external threats. *Segmenting* the network into smaller parts can reduce the impact of a security breach, whereas *network security* entails protecting the network infrastructure, such as routers, switches, and firewalls.

11.4.3   End Point Protection and Security Monitoring

End point protection and security monitoring are essential cybersecurity controls for intelligent and sustainable manufacturing systems. *End point protection* involves securing the end points, such as desktops, laptops, and mobile devices, while *security monitoring* involves monitoring the system for security threats and anomalies.

11.4.4 INCIDENT RESPONSE AND DISASTER RECOVERY

Critical cybersecurity measures like incident response and disaster recovery serve to lessen the effects of security incidents and maintain company continuity. *Disaster recovery* comprises overcoming a security incident and returning the system to normal operations, whereas *incident response* entails promptly and effectively responding to security incidents. These measures aid in making the system resilient to security incidents and capable of swift recovery from any disturbance.

11.5 TRAINING AND AWARENESS FOR MANUFACTURING PERSONNEL

11.5.1 TRAINING ON CYBERSECURITY POLICIES AND PROCEDURES

Training manufacturing personnel on cybersecurity policies and procedures is crucial in mitigating cyber threats. The training should cover areas such as password management, network security, data backup and recovery, phishing attacks, and malware protection. It is essential to provide regular training to keep employees informed about new cybersecurity threats and how to respond to them.

11.5.2 DEVELOPING A CULTURE OF CYBERSECURITY AWARENESS

Developing a culture of cybersecurity awareness within the manufacturing organization is essential. Employees should be encouraged to report any suspicious activity or potential cyber threats immediately. This culture can be established by providing regular cybersecurity training, creating cybersecurity awareness campaigns, and promoting a sense of responsibility among employees for the security of the organization's data.

11.5.3 REGULAR TESTING AND SIMULATION OF CYBERATTACKS

Regular testing and simulation of cyberattacks help identify vulnerabilities and assess the effectiveness of cybersecurity measures. The manufacturing organization can conduct penetration testing, phishing simulation exercises, and vulnerability scans to identify weaknesses in the system. The testing results can be used to enhance the organization's cybersecurity policies and procedures and to ensure that employees are adequately trained to respond to cyber threats.

11.6 CONTINUOUS IMPROVEMENT AND ADAPTATION

11.6.1 REGULAR ASSESSMENT OF CYBERSECURITY RISKS

Any cybersecurity program must include a regular assessment of cybersecurity threats. This entails identifying potential risks and weaknesses to the organization's assets, such as its data, networks, and information systems, and assessing the likelihood and potential consequences of these risks.

11.6.2   MONITORING OF EMERGING THREATS AND VULNERABILITIES

In addition to regularly assessing cybersecurity risks, it is also crucial to monitor emerging threats and vulnerabilities. This involves keeping up-to-date with the latest trends and developments in the cybersecurity landscape, such as new types of malware or hacking techniques, and taking appropriate action to mitigate these risks.

11.6.3   UPDATING AND IMPROVING CYBERSECURITY CONTROLS AND POLICIES

As new threats and vulnerabilities emerge, it is important to update and improve cybersecurity controls and policies to ensure that they remain effective. This entails routinely evaluating and testing security controls, such as firewalls and intrusion detection systems, as well as changing policies and procedures to take into account changes in the threat environment and regulatory requirements.

11.6.4   IMPLEMENTING A CONTINUOUS IMPROVEMENT PROCESS

Finally, to ensure ongoing effectiveness and relevance of the cybersecurity program, it is important to implement a continuous improvement process. This involves regularly reviewing and refining the program based on feedback and performance metrics, identifying areas for improvement, and implementing changes to address these issues. By continuously improving and adapting the cybersecurity program, organizations can better protect their assets from cyber threats.

Intelligent and sustainable manufacturing has the potential to revolutionize manufacturing processes, improve efficiency, and reduce environmental impact. However, the integration of smart technologies into manufacturing processes presents new cybersecurity risks that need to be addressed. By implementing best practices for cybersecurity, businesses can ensure that their systems are secure and their operations are not disrupted. As the adoption of intelligent and sustainable manufacturing continues to grow, it is essential to prioritize cybersecurity to ensure that the benefits of these technologies are realized without compromising security.

The results of prioritizing cybersecurity in intelligent and sustainable manufacturing can be significant. By implementing best practices for cybersecurity, businesses can reduce the risk of cyberattacks, protect sensitive information, and ensure that their operations run smoothly. The following are some of the possible results of effective cybersecurity measures in intelligent and sustainable manufacturing:

- *Reduced risk of cyberattacks.* By identifying potential vulnerabilities and implementing appropriate cybersecurity measures, businesses can significantly reduce the risk of cyberattacks.
- *Protection of sensitive data.* Encryption, authentication protocols, and other cybersecurity measures can help protect sensitive data from unauthorized access.

- *Improved operational efficiency.* Effective cybersecurity measures can help prevent disruptions to operations caused by cyberattacks, ensuring that production continues without interruption.
- *Enhanced reputation.* Prioritizing cybersecurity and implementing robust measures can help businesses build a reputation for reliability and trustworthiness.
- *Compliance with laws.* There are laws in many nations requiring organizations to take the proper cybersecurity precautions. Businesses can stay out of trouble and avoid fines by adhering to these rules.

11.7   RESULTS AND DISCUSSION

The proposed method demonstrates a more comprehensive and proactive approach to security architecture, design principles, integration in the SDLC, compliance with standards/regulations, and overall cybersecurity. It addresses the limitations of existing methods, which often prioritize perimeter-based security and reactive measures. By considering the strengths of the proposed method, organizations can enhance their cybersecurity posture and better protect their systems and data from emerging threats, as shown in Table 11.1

The proposed method demonstrates a more comprehensive and robust approach to various security aspects compared to existing methods, as shown in Table 11.2. It emphasizes defense in depth, secure communication, least privilege, secure coding, threat intelligence, incident response, and compliance with cybersecurity standards. By considering the strengths of the proposed method, organizations can enhance

TABLE 11.1

A Comparative Analysis of Secure Architecture and Integration Approaches in Proposed and Existing Methods

Methods/Features	Proposed Method	Existing Methods
Secure architecture	Emphasizes defense in depth	Primarily focuses on perimeter-based
Design principles	Separation of duties, least privilege, secure communication	Security measures like firewalls and access controls
Integration in SDLC	Incorporates cybersecurity considerations at every stage	May lack comprehensive approach in early stages of system design
Compliance with standards/ regulations	Highlights the importance of adhering to relevant cybersecurity standards and regulations	Some methods focus on meeting regulatory compliance requirements without considering full cybersecurity risks and best practices
Proactive approach	Emphasizes proactive security measures during system design	Some methods rely heavily on incident response and disaster recovery

TABLE 11.2

A Comparative Analysis of Security Aspects in the Proposed Method and Existing Methods

Security Aspects	Proposed Method	Existing Methods
Defense in depth	Implements a layered security approach, with multiple security controls at various levels	May rely on single or limited security measures, such as firewalls and access controls, without a layered approach
Secure communication	Prioritizes secure communication protocols and encryption techniques to protect data in transit	May overlook the importance of secure communication and rely on insecure protocols or lack encryption
Least privilege	Adheres to the principle of least privilege, granting users only the necessary permissions	May not implement strict least privilege access controls, resulting in broader access rights
Secure coding	Encourages secure coding practices to prevent common vulnerabilities, like injection attacks	May lack emphasis on secure coding practices, leading to potential vulnerabilities and code exploits
Threat intelligence	Incorporates threat intelligence to identify and mitigate potential security risks and attacks	May not actively monitor and analyze emerging threats or have a proactive approach toward threat intelligence
Incident response	Establishes a structured incident response process to detect, respond to, and recover from security	May not have a well-defined incident response plan or a comprehensive process for handling security incidents
Compliance	Focuses on adherence to relevant cybersecurity standards and regulations for comprehensive security	Compliance efforts may be limited to meeting regulatory requirements without considering full cybersecurity risks and best practices

their overall security posture and better protect their systems and data from potential threats and vulnerabilities.

Table 11.3, titled "Comparative Analysis of Performance Metrics, Security Incidents, and User Feedback in Intelligent and Sustainable Manufacturing Systems: Proposed Method vs. Existing Method," provides a comprehensive comparison between the proposed method and an existing method in the context of intelligent and sustainable manufacturing systems. The purpose of the table is to evaluate the performance, security incidents, and user feedback of both methods and identify their strengths and weaknesses. This analysis is crucial for understanding the effectiveness and suitability of each method in real-world manufacturing environments. In terms of performance metrics, the table compares various aspects, such as throughput, cycle time, efficiency, and resource utilization. These metrics provide insights into the overall efficiency and effectiveness of the manufacturing system under each method. Table 11.3 also addresses security incidents, including the types of incidents, the frequency, and the severity. It evaluates how well each method handles security

TABLE 11.3

Comparative Analysis of Performance Metrics, Security Incidents, and User Feedback in Intelligent and Sustainable Manufacturing Systems: Proposed Method vs. Existing Method

Analysis Aspect	Performance Metrics	Security Incidents	User Feedback
Measurement method	Quantitative data	Incident reports, tracking systems, and security audits	Surveys, interviews, feedback forms
Statistical techniques	Descriptive statistics, trend analysis, regression analysis	Frequency counts, correlation analysis	Thematic analysis, sentiment analysis
Proposed method algorithm	Machine learning algorithms (e.g., decision trees, random forests, neural networks)	Anomaly detection algorithms (e.g., clustering algorithms, statistical models, rule-based systems)	Natural language processing algorithms (e.g., text classification algorithms, sentiment analysis algorithms)
Existing method algorithm	Statistical process control (SPC) techniques (e.g., control charts, process capability analysis)	Intrusion detection systems	Qualitative coding algorithms, thematic analysis algorithms, interpretative analysis algorithms

vulnerabilities, threats, and attacks, and the level of protection they offer to ensure the integrity and confidentiality of the manufacturing system and its data.

11.8 CONCLUSION

Cybersecurity is a critical aspect of intelligent and sustainable manufacturing, and it is essential for ensuring the stability, reliability, and security of these systems. By implementing robust cybersecurity measures, manufacturers can minimize the risk of disruptive incidents, protect their data and systems, and build trust with their stakeholders. Integrating cybersecurity into the design and implementation of intelligent and sustainable manufacturing systems is crucial to ensure the protection of critical data, intellectual property, and manufacturing processes.

REFERENCES

1. D. Zhang, X. Guo, and H. Liu, "Cybersecurity and privacy protection in intelligent manufacturing systems: Challenges and solutions," *Journal of Intelligent Manufacturing*, vol. 31, no. 7, pp. 1701–1704, 2020.
2. F. Deng, W. He, Y. Ren, and Y. Sun, "A survey of security and privacy issues in Industry 4.0," *Journal of Network and Computer Applications*, vol. 136, pp. 1–10, 2019.
3. L. Bai, Z. Li, Y. Li, and X. Li, "Cybersecurity issues in intelligent manufacturing: A review," *Journal of Intelligent Manufacturing*, vol. 31, no. 7, pp. 1765–178, 2020.

4. R. Huang, J. Chen, and W. Dong, "A survey of cybersecurity for smart manufacturing," *International Journal of Production Research*, vol. 57, no. 24, pp. 7616–7632, 2019.

5. S. Han, Y. Zhao, C. Guo, and Y. Zhang, "A survey on security in intelligent manufacturing," *Journal of Advanced Mechanical Engineering*, vol. 12, no. 4, pp. 1–14, 2020.

6. W. Huang, W. Ren, and Q. Wang, "Cybersecurity challenges and opportunities in intelligent manufacturing: A review," *IEEE Access*, vol. 7, pp. 27082–27093, 2019.

7. W. Sun, Y. Hu, X. Zhang, and K. Wang, "A review of cybersecurity challenges in smart manufacturing," *Journal of Cleaner Production*, vol. 284, no. 1, p. 124705, 2021.

8. W. Zhang, W. Li, Y. Li, and Y. Li, "Cybersecurity for intelligent manufacturing systems: A review," *International Journal of Advanced Manufacturing Technology*, vol. 107, no. 1–2, pp. 225–240, 2020.

9. X. Wu, H. Yu, Y. Zhang, and J. Sun, "A survey of cybersecurity in the age of intelligent manufacturing," *Journal of Network and Computer Applications*, vol. 143, pp. 1–13, 2019.

10. Y. Li, L. Yang, X. Wang, and W. Sun, "A review of cybersecurity in cyber-physical systems and its implications for intelligent manufacturing," *Journal of Intelligent Manufacturing*, vol. 31, no. 7, pp. 1755–1773, 2020.

11. Y. Zhou, M. Shojafar, B. B. Gupta, and M. Conti, "A review of cyber security risks and defences in industrial control systems," *Journal of Network and Computer Applications*, vol. 98, pp. 81–102, 2017.

12. Z. Wang, X. Wang, W. Sun, and L. Yang, "Cybersecurity challenges and solutions in intelligent manufacturing: A review," *Journal of Ambient Intelligence and Humanized Computing*, vol. 11, no. 7, pp. 3085–3097, 2020.

13. Z. Zhou , L. Zhang, Y. Sun, J. Qian, and H. Wang, "A review of cybersecurity research in intelligent manufacturing systems," *Journal of Intelligent Manufacturing*, vol. 31, no. 7, pp. 1705–1717, 2020.

12 An Invisible Threat to the Security of Nations in the Age of "Deepfakes"

Divyansh Shukla and Anshul Pandey

12.1 INTRODUCTION

"If you think technology can solve your security problems, then you don't understand the problems and you don't understand the technology."
—Bruce Schneier

If we closely examine this quote by the American cryptographer Bruce Schneier, it reveals the current security problems of India and, in general, the security problems of any country in the world. With the advancement of technology in every sphere of life, it leads to a number of invisible problems. One such problem that is knocking on the doors of every jurisdiction, due to technological advancement, is the issue of threat to sovereignty and the security of the state. India is also facing the challenges posed by the emerging arms of the technological capabilities of AI that produce deepfakes. Before digging deeper into the invisible threat of deepfakes, we must have to first analyze what *sovereignty* means in context with the security of the nation.

Sovereignty, which refers to the supreme will of the state, gets threatened when other state or non-state actors attempt to get control over the subjects and internal affairs of the state. [1] As per the well-recognized international law principle, the idea of sovereignty is applicable to cyberspace [2] as well [3]. Therefore, the unauthorized access to information and communication technology (ICT) abroad by third parties without the knowledge or consent of the host nation and/or its law enforcement agencies can be a breach of state sovereignty. As per the United States International Strategy for Cyberspace, actions like cyberattacks, cyber exploitation, and other hostile acts in cyberspace that jeopardize peace, stability, civil rights, and privacy are seen as territorial sovereignty violations [4]. This chapter is anchored on the concept that sovereignty and security need to be understood as two sides of the same coin. Threat to security, in turn, will affect territorial sovereignty, as the state becomes incapable of exercising its will, and this leads to the collapse of the whole scheme of, say, the Constitution of India due to the dilution of the basic structure of the Indian Constitution. The Honorable Supreme Court of India has observed in *Kesavananda Bharti v. State of Kerala* [5] that sovereignty is the basic feature of the Indian Constitution.

DOI: 10.1201/9781003501152-12

Securing the nation not only through its territorial borders but also from techno-logical weapons capable of violating the cyberspace is essential, as it directly attacks the core values of our holy Constitution. This chapter will discuss how deepfakes pose a new and emerging threat to the security of India, and in the latter part, it will analyze whether the present legislations are well equipped to deal with the emerging and devastating impacts of deepfakes in the context of security of India. It will also analyze the role of the Indian Computer Emergency Response Team (CERT-In) in dealing with the invisible threat of deepfakes, and it will also offer some suggestions to lawmakers of India on how India will have to respond and enhance its security to tackle deepfakes through the counter-weapon of law.

12.2 WHAT ARE DEEPFAKES?

Emerging technologies are enablers of a better future; they are these empower-ing notions which hold the power to dictate the future of humanity. The advent of artificial intelligence (further referred to as AI) has given birth to numerous other branches of technologies through advancing data sciences. Deepfakes are part of the family tree and are considered as part of synthetic media. *Deepfake* is a method of technology which uses AI as a means whereby the user has the ability to recreate through audiovisual cues a synthetically augmented video of a real person, including pictures, by making them act or speak things that have not been committed in reality. The algorithms and systems hold the ability, through the means of machine learning, to process the collection of data and information for the user so that any form of body part, including the face, body, and other visual attributes, can be regenerated on-screen and appear seemingly real but not in reality [6].

Deepfakes are a derivative of deep learning, which is under the heading of AI, and the technology works on the basis of neural networks. The neural network tech-nology in deep learning is frequently seen to be filled with input/output structures. The algorithm consists of two related bodies which are known as the generator and the discriminator. These algorithms are very significant because they distinguish the content between fake and real. The generator set is used to create the fake content, while the discriminator set is used to distinguish the features which were faked, hence authenticating the material. After the detection of the authentic features within its system, the discriminator reports it back to the generator so that the fake content can be perfected more and more and be in line with the real instance. So the system improves itself through such information. The input function gets more weightage when the picture or content is closer to the real image; essentially, it is like a scale of success to determine the degree of correctness.

The underlying technology can overlay face images, create facial motions, switch faces, maneuver facial expressions, produce faces, and synthesize the speech of a target individual onto a video of a spokesperson in order to create a video of the target individual acting similarly to the source person. The subse-quent impersonation is often practically indistinguishable from the original ones [7]. Videos of Barack Obama [8], Donald Trump [9], Nancy Pelosi [10], Russian president Vladimir Putin [11], Ukrainian president Volodymyr Zelenskyy [12], the economic affairs minister of Malaysia [13], Tom Cruise [14], Facebook CEO Mark Zuckerberg [15], American president Richard Nixon [16], and Queen Elizabeth [17] are a few deepfake incidents that reflect the endless creativity of this technology. India, for the first time, witnessed the emergence of deepfake manipulation in the assembly election of New Delhi in the year 2020, when a deepfake video of Manoj Tiwari, the state president of Bhartiya Janta Party (BJP), was widely circulated through WhatsApp [18]. Since then, adult deepfakes of Indian celebrities have been on the rise, [19] and very recently, famous Indian actor Anil Kapoor has successfully filed a suit for interim injunction against the defendant to seek protection of various other attributes of his personality against misuse of all hues over the internet. The actor argued that the defendant falsely endorsed him as a motivational speaker [20].

False videos showing corrupt authorities, atrocities committed by the military, immoral presidential candidates, and emergency professionals warning of a terrorist attack can be produced with the aid of AI technology [21]. The diverse domains in which harm can be caused by deepfake videos or images can be best understood from the Deepfakes Accountability Act [22]. The Act imposes criminal liability if the advanced technology is used to create false personation with the intent to:

1. Feature a person in sexual activity
2. Incite physical harm or violence
3. Incite armed or diplomatic war
4. Obstruct an official action
5. Engage in deception, such as securities fraud
6. Influence the discussion of a domestic policy
7. Tamper with the election in a territory, state, or federal government

The threat of deepfakes is strengthened by the fact that they are extraordinarily precise, are easy to create, and have adverse effect on the viewers. Moreover, the quality of deepfake video seeming to be real will keep improving over time [23], to the point that it will be difficult to detect fakes for unaided humans, and as a result of this, it will be difficult for people and AI systems themselves to differentiate real videos from fake ones [24]. This threat makes the detection of deepfakes a continuing problem.

Governments all around the world, the tech industry, [25] and other stakeholders have made efforts to develop the technology for the detection of deepfakes. The main goal of deepfake detection is to figure out the authenticity of video recordings and to find out if the video has been manipulated in any manner or not.

Technology is an enabler for a better future. It has contributed, more than anything, to making the life of humans "a life of convenience and luxury." Each and every field, in the life of humans, owes its development to technology. On the other hand, it has also witnessed a lot of landmark evolution, which also makes it an invisible danger to mankind. Deepfake, which is part of deep learning technology, is one such technological development.

12.3 DEVASTATING AND INVISIBLE IMPACT OF DEEPFAKES ON SECURITY AND SOVEREIGNTY

The arms of deepfake are so big that the wrongs associated with it are indeterminable and unanticipated. One of the most often cited risks associated with deepfakes is the potential for disinformation and other internet manipulation to be facilitated and amplified, thus weakening the fundamental framework of the Indian Constitution, such as threats to democracy, sovereignty, integrity, and fraternity. Disinformation is by no means a new issue, but in the current global context, it poses a difficulty due to recent technological advancements related to AI-generated deepfakes, which have boosted the manipulative potential of video- and audio-based contents. This chapter will discuss various instances where disinformation had negatively affected the principle of sovereignty globally due to the evil effects of deepfakes.

12.3.1 DEEPFAKE AUDIO OR VIDEO THAT SHOWS RACIST, ABUSIVE, ANTI-RELIGIOUS, AND VIOLENT COMMENTS BY A POLITICAL LEADER LEADING TO DISTURBANCE OF INTERNAL PEACE

India is a country with a lot of diversity. Our Constitution seeks to prevent any erratic inclinations toward strengthening Indian unity via assimilation of diversity [26]. It is essential to state that India represents a social, religious, and cultural diversity. Unity in diversity is the Indian culture and ethos [27]. By noticing the impact of deepfakes in the last couple of years, it can conclusively be said that deepfakes have the full potential to attack this fabric of "unity in diversity," and when this fabric is attacked, law and order, as well as public order, will be at its lowest level and the whole scheme of the Constitution of India will collapse. Some of the examples which are noticeable are as follows:

1. On January 29, 2023, a number of 4chan users created voice deepfakes of well-known people, such Emma Watson, Joe Rogan, and Ben Shapiro, by using ElevenLabs, a "speech synthesis" and "voice cloning" service. These deepfake recordings featured violent, insulting, and racial remarks [28].
2. Russia has previously attempted to inflame tensions around Black Lives Matter demonstrations and other protests by spreading false information to divide American society. Deepfakes that depict White police officers shooting down unarmed Black men while using racist epithets can heighten the tension. Such initiatives can deepen rifts among the armed forces and erode public trust in political figures [29].
3. In 2020, Manoj Tiwari, the president of the ruling Bhartiya Janata Party (BJP) of India, was the target of a deepfake that spread across the nation's WhatsApp users [30]. On February 7, a number of videos appeared on WhatsApp in which Tiwari stated, "Kejriwal stated that he would open 500 new schools. Have they gotten underway yet? He declared that 15 lakh CCTV cameras would be installed. Are they installed? On the basis of these

kinds of assurances, he simply conned us. Delhi now has an opportunity to make a complete difference. Kindly click the Lotus button on February 8 to establish a government headed by Modi Ji" [31].

These examples depict the potential harm that deepfakes may pose to the integrity of India. If a deepfake audio or video making racist, abusive, anti-religious, and/or violent comments of a famous personality or an army officer or a political leader or a constitutional post-holder of India is created and circulated in the public, then there are high chances that public peace may get jeopardized. The people of India may start agitations, go on strikes, or may choose other violent forms to express their anguish due to such deepfake video. Furthermore, these deepfakes can exacerbate rifts among the armed forces and undermine trust in political authorities. These incidents become a major invisible threat to the security and sovereignty of India. The external power, while taking advantage of these riots or divisions within military or the low confidence on political leaders, can increase intrusion in India. An example of a deepfake that may legitimize or encourage an uprising was provided by Robert Chesney and Danielle Citron [32]. The video purportedly showed a US general burning a Koran. Such bogus movies in India could add gasoline to the fire at a time when polarization has been progressively rising due to a divided political leadership, India's economic transition, media landscape changes, and the emergence of competitive caste politics [33].

12.3.2 CYBERATTACKS

Deepfakes are becoming a more common weapon used by cybercriminals in their attacks. In August 2022, VMware announced the findings of a survey for its eighth annual Global Incident Response Threat Report [34], which revealed that 66% of cybersecurity professionals surveyed have witnessed the use of deepfakes in a cyberattack. That is a yearly gain of 13%. It was stated that 78% of those attacks were delivered by email.

12.3.3 DEEPFAKE RANSOMWARE

Another star has been added to the cyberattack scene by the deepfake ransomware. Paul Andrei Brickman first used the term "deepfake ransomware" in public. This kind of malicious software creates fictitious videos automatically, depicting the victim in an intimate or incriminating scene, and then threatens to share them unless a ransom is paid [35]. This elevates sextortion to a dangerous degree and could be lethal. A threat actor films a deepfake of their intended victim, captures screenshots of this video, and assuming the identity of a trustworthy contact, emails the screenshots and a link to the purported video to the intended recipient so that, in case of doubt, they can view it on their own. The target, intrigued and possibly a little afraid, clicks the link, is sent to a brief video displaying themself in a compromised position, and while this is happening, ransomware is being downloaded onto their computer. Alternatively, the URL can actually download and run a ransomware file instead of a supposed film.

12.3.4 CYBERBULLYING

The United States Department of Homeland Security provided an example of this kind of hack. A deepfake of a person is created to show a target in a circumstance that could harm his or her reputation or limit their ability to access organizations, resources, or advantages; this could involve showing the person engaging in illegal activity. The attacker wants to damage the victim's reputation, which could also serve to elevate the standing of a different target that they favor. In a well-reported instance that occurred recently in Pennsylvania, a mother tried to harm the reputations of her daughter's friends who were vying for a small number of slots on a cheerleading squad [36]. In this case, a deepfake video showing the target acting criminally is created and distributed to people who have control over the target's actions. These authorities prohibit or take away the target's ability to engage in particular activities based on the video.

12.3.5 PHISHING

Numerous instances of deepfake-enabled phishing have been reported already:

1. The insurance company Euler Hermes, who paid for the event, disclosed the details of the first known deepfake attack, which happened in March 2019. The CEO of a UK energy company received a call from someone he believed to be his boss, the head of the company's German parent company, which was how the fraud got its start. Euler Hermes claims that the CEO of the United Kingdom heard his employer's tone, intonation, and faint German accent quite right. The boss asked him to transfer $243,000, purportedly into the account of a Hungarian supplier. The CEO of the energy company complied with the request, only to discover afterward that he had been duped. This, according to the insurance company's fraud experts, was an instance of an AI-driven deepfake phishing attack.
2. Fraudsters employed a deepfake hologram of a chief communications officer for cryptocurrencies on Zoom calls in an attempt to fool executives into divulging private information.
3. Using real-time voice cloning, thieves were able to mimic the voice of the director of a Dubai bank and trick the manager of a Hong Kong bank into sending $35 million to their group. The manager "recognized it" over the phone because the false voice sounded so realistic.
4. An organization believed that they had employed a remote worker to handle technical support. Instead, they employed a criminal who used deepfake technology to construct a phony persona and steal personally identifying information in order to access the company's network and data.
5. Hackers impersonating a CEO left phony voicemails demanding that staff members and outside vendors donate to humanitarian or disaster relief organizations or make investments through fictitious websites that diverted money to offshore accounts [37].

Deepfakes can be used to trick targets into providing personal information, account credentials, or money. Now, just consider an incident in the case of India. If a person who is working in an institute of national importance becomes the target of such cyberattacks mentioned earlier, then there are high chances that the security of the cyber infrastructure, databases, confidential files, etc. may get compromised, and the threat actor may gain unauthorized access to such data, leading to threats to the security of India, which, in turn, will affect the sovereignty of India.

12.4 INTERNATIONAL CONFLICTS

Although it can be argued that deepfake videos will eventually be discovered to be false, until that time, it can have devastating impacts, especially in matters of international affairs.

12.4.1 FALSIFYING ORDER

Daniel L. Byman, Chongyang Gao, Chris Meserole, and V. S. Subrahmanian argue that through deepfakes, one can make a video or audio falsifying an order, and one can also divide allies in a war. A classic example justifying their argument is given thus: Zelenskyy was seen on Russian video ordering Ukrainian soldiers to lay down their weapons and submit to the advancing Russian army. Some fictitious order might appear in videos of senior leaders ordering soldiers to lay down their guns, retreat, use fake chemical weapons, or urge mass surrenders in order to remove well-defended soldiers or weaken the force.

12.4.2 SOWING CONFUSION

When soldiers and civilians are told to disregard orders from leaders as possible forgeries, they could unintentionally disregard genuine commands as well, leading to confusion during a potentially hazardous situation. When fiction and reality are mixed together to make it impossible for people to tell them apart, disinformation campaigns work best.

The president of Gabon was reportedly deepfaked to give a stiff, emotionless speech, which made his detractors doubt his capacity to lead, and the military made an effort at overthrowing the government. Since it was unclear if the president was ill or whether the speech was a deepfake in this instance, various actors may have come to different conclusions about whether the president was still fit to hold office as a result of the ambiguity.

This kind of confusion is particularly possible when the deepfake feeds back into pre-existing cognitive biases that make it difficult to accept unsettling information, such as proof that a well-liked leader is doing inappropriately. Similar to this, news outlets could be reluctant to cover breaking news out of concern that they would fall for a deepfake. As Chesney and Citron note, the potential for deepfakes does, in fact, produce a "liar's dividend," enabling political figures to contest the veracity of their own sincere transgressions. This unintentional misunderstanding is essentially the opposite of misleading orders in that valid instructions are ignored rather than being followed [38].

12.4.3 DISCREDITING LEADERS

Deepfakes provide a potentially convincing way to portray leaders acting in a way that would damage a bilateral or multilateral relationship, such as making disparaging remarks about allies and allied casualties; making fun of something that is important to an ally, like the flow of refugees or energy shortages; or acting in other ways.

In Myanmar, a suspected deepfake was used to portray a former minister of state as having bribed Aung San Suu Kyi, the country's former leader and a target of the military authority. Deepfakes may have been used more skillfully to create the staged video of Speaker of the House of Representatives Nancy Pelosi who was purportedly intoxicated. Deepfake films may depict a leader acting in ways that outrage significant nations, peoples, and constituencies, or uttering racist, harsh, or insensitive things while sneering at victims or political allies.

The aforementioned scenarios show that one party to a war can make a deepfake video of another party which depicts that the latter party is involved in anti-national activities. If the Chinese or Pakistani army created a deepfake video of the Indian army which shows the Indian army involved in anti-Chinese or anti-Pakistani activities, then immediate impacts could be devastating. They can justify their illegal actions on the borders in the shelter of these deepfake videos. In the year 2019, on September 11, a Twitter user going by the handle "Eva Zheng" said that Indian troops had infiltrated and were monitoring a whole region that was under Chinese sovereignty [39]. In recent years, there have been multiple instances of the Pakistani army creating fake Twitter profiles of Indian army generals to spread fake stories. These stories are then used as justifications for the incursions.

The deepfake technology thus presents a huge problem for the Indian government and the army if enemy states start using it to morph videos and spread them around to defame India or use it as a pretext to take aggressive actions [40].

12.5 A LOOK INTO THE WAR BETWEEN DEEPFAKES AND INDIAN LEGISLATIONS

In the preceding section, we have analyzed the harms which deepfakes pose to the security or sovereignty or integrity of nations all around the world. In this section, we are going to analyze whether present legislations can be invoked against deepfake creators and, if yes, whether provisions are well equipped or not to deal with this threat.

12.5.1 NATIONAL SECURITY ACT

The National Security Act 1980 (hereafter, the "Act") states in its preamble that its goal is to establish preventive detention in specific circumstances and to address related issues [41]. This Act generally lays down the cases in which the preventive detention of a person can be done. The provisions of this Act are wide enough to be invoked against any person involved in the act of disturbance of public order. This Act is not only applicable to terrorists but also to any act or omission which has the potential to disturb public order or the security of the state or of India. The Supreme Court has mentioned that "[a]n act is deemed disruptive to the upkeep of public order if it results in the disruption of the steady pace of life within a community or a designated area" [42].

A fake video was shared on WhatsApp, and it resulted in violence within different communities, in which 62 people were killed and 50,000 got displaced from their homes in Uttar Pradesh [43]. If we look at the survey conducted by Microsoft, we find that 64% of Indians encounter fake news daily, and it is highest among all the countries surveyed. Fake news has the potential to incite violence, and any act which incites or tends to incite violence is also going to disturb public order, and any act which is prejudicial to the maintenance of public order will be in the domain of the National Security Act.

In *Tehseen S. Poonawalla v. Union of India*, the Honorable Supreme Court acknowledged that lynching and mob violence are creeping threats that can eventually take the form of a typhon-like monster. This is demonstrated by the growing wave of incidents of recurring patterns by frenzied mobs across the nation incited by intolerance and misinformed by the circulation of fake news and false stories.

As deepfakes are a sophisticated form of creating realistic fake video or audio, it therefore eventually has the potential to incite violence, which may lead to overall disturbance of public order. This leads to the application of the National Security Act over deepfake creators also. Spreading fake news or disinformation is nothing but an act of abetment of disturbance of public order with mala fide intentions.

The central government or state government may issue an order for the detention of an individual under Sub-Sections 1 and 2 of Section 3 of this Act if they are convinced that the individual may act in a way that jeopardizes India's defense, its relations with foreign powers, the security of the state, or the upholding of public order.

If a deepfake video or audio is created making racial, abusive, anti-religious, or violent comments, or if any deepfake audio or video of any notable person is created which is threating the security of India in any manner defined in the article so far, by any person, then the government has the sufficient cause to be satisfied under Section 3 of the Act that such person is a threat to the security of India, and it can then pass an order to detain that person. If we further look at Sub-Section 3 of Section 3, then in our opinion, this provision can also be used as a weapon against deepfake creators. Once a deepfake audio or video is created, there are more chances that the security, integrity, or sovereignty of India, or the maintenance of public order of a state or district, come into question, and certain circumstances that are not in the interest of the public may prevail. The term "likely to prevail" under Sub-Section 3 of Section 3 is so wide that it covers all the aforesaid conditions and the concerned person can be detained. The essence of Sub-Section 3 of Section 3 lies in the mere fact that it can be used to control deepfakes at the ground level as authorization may come from the district magistrate or the commissioner of the police to detain deepfake creators. It will also set an example for those who are thinking of using this technological weapon against the very idea of the Constitution.

12.5.2 INFORMATION TECHNOLOGY ACT

The Information Technology Act 2000 (hereinafter, the IT Act) currently is the primary legislation that can be used to tackle the immense threat that deepfakes pose to the security, integrity, or sovereignty of India. Deepfake creators can be charged under the various sections of the IT Act.

A person's reputation and credibility may suffer if deepfakes are used to steal someone's identity, fabricate false information about them, or sway public opinion. Deepfakes can also distribute incorrect information. These offences are punishable under Sections 66 (computer-related offences) and 66-C (punishment for identity theft) of the Information Technology Act 2000 [44].

Deepfakes are a severe problem that can have long-lasting effects on society if they are used to propagate false information, undermine the government, or stir up hatred and discontent against it. Spreading false or misleading information has the potential to weaken public trust, confuse the public, and be used to sway political outcomes or manipulate public opinion. These offences are punishable under the Information Technology Act of 2000's Section 66-F (cyberterrorism) and the Information Technology (Intermediary Guidelines and Digital Media Ethics Code) Amendment Rules 2022.

Deepfake-based internet defamation and hate speech are important problems that can be harmful to both people and society as a whole. These crimes are punishable under the Information Technology Act of 2000's Information Technology (Intermediary Guidelines and Digital Media Ethics Code) Amendment Rules 2022. Deepfakes in elections have the potential to have serious repercussions and compromise the democratic process's credibility. These offences are punishable under Sections 66-D (penalty for cheating by personation by utilizing computer resource) and 66-F (cyberterrorism) of the Information Technology Act 2000.

This technology has the ability to create fake images or videos of people performing or saying things that have never really happened, which could damage people's reputations or spread false information. The Information Technology Act of 2000 contains provisions for the prosecution of the following crimes: Section 66-E, which deals with privacy violations; Section 67, which deals with publishing or transmitting obscene material in electronic form; Section 67-A, which deals with publishing or transmitting material containing sexually explicit acts, etc., in electronic form; and Section 67-B, which deals with publishing or transmitting material depicting children in a sexually explicit act/pornography in electronic form.

12.6 COMPARISON WITH OTHER MAJOR NATIONS

Despite all these provisions, in our opinion, Indian law lags behind when it comes to tackling the threat that deepfakes pose to the security of India. If we look at the global scenario, certain countries have passed specific or special law that regulates the unwanted and malicious use of this advanced technology.

12.6.1 UNITED STATES OF AMERICA

The Deepfake Accountability Act [45] proposed in the US Congress imposes criminal liability if the advance technology is used in any of the following ways:

1. Featuring person in sexual activity
2. Causing violence or physical harm
3. Inciting armed or diplomatic conflict
4. Interfering in an official proceeding
5. Committing fraud, including securities fraud
6. Influencing a domestic policy debate
7. Interfering in a federal, domestic, or territorial election

With DARPA's Media Forensics (MediFor) program, the Department of Defense is attempting to combat deepfakes [46]. DARPA researchers are working on technologies under MediFor that will be able to determine automatically if a picture or a video has been manipulated. The program's objective is to provide an end-to-end media forensics platform that can identify modifications and provide information on how they were carried out.

Furthermore, for the next five years, the Department of Homeland Security (DHS) must publish an annual report on deepfakes in accordance with the National Defense Authorization Act 2021. Any possible harm from the technology should be covered in the report, including harm against certain demographics, fraud, and foreign influence efforts [47]. Furthermore, the law directs DHS to research technologies for creating deepfakes, as well as the potential means of detection and mitigation. Lastly, the law mandates that the US Department of Defense research the potential for enemies to produce deepfake material that features members of the armed forces or their families and suggest modifications to existing policies.

The Identifying Outputs of Generative Adversarial Networks Act 2020 mandates that the National Institute of Standards and Technology support the development of deepfake standards, that the National Science Foundation conduct research on deepfake technology and authenticity measures, and that both agencies devise strategies for collaborating with the private sector on deepfake identification capabilities.

Virginia has criminalized the distribution of non-consensual deepfake images and videos [48]. California passed two laws in September 2019 which will be effective from 2020 onward which aims to regulate the distribution of such deepfake [49]: (1) California Bill AB 730 prohibits the use of deepfake in election campaigns. (However, this bill is with a sunset clause and thus will not be effective until January 1, 2023). (2) California Bill AB 602 recognizes deepfakes and pornography. (This bill does not have the sunset clause.) Till now, it is clear that detecting deepfake is tackling deepfake, and the most important thing is, the detection of deepfakes must be done as soon as possible; otherwise, it will be too late, and till then, the harm that deepfake causes is incalculable.

12.6.2 China

This year on January 10, China's Cyberspace Administration (CAC) unveiled new regulations to deal with deepfakes. The main essence of this legislation is the regulation of the threat of deepfakes from their root cause, that is, content generation service providers, and this aspect makes it a more comprehensive mechanism to preserve social stability. Deep synthesis (also known as deepfake) technology and services, such as text, images, audio, and video generated using AI-based models, are regulated under a new law in China called the Deep Synthesis Provisions. Given China's lengthy history of striving to maintain strong control over the internet, these new laws are hardly shocking [50]. The Cyberspace Administration of China (CAC) claims that deep synthesis technology has been "used by some unscrupulous people to produce, copy, publish, and disseminate illegal and harmful information, to slander and belittle others' reputation and honour, and to counterfeit others' identities," and that as a result, regulations are necessary. The CAC continues, "Committing fraud, etc., affects social order and communication, harms people's legitimate rights and interests, and puts social stability and national security at risk." Both platform providers that offer content generation services and end users that utilize those services are subject to rules and regulations. A watermark, which is a piece of text or an image that is visibly put on a video to indicate that the content has been modified, is required by these new Chinese regulations for any content produced using an AI system. The establishment of feedback mechanisms for content consumers and other rules, such as the evaluation and verification of AI algorithms deployed and user authentication (to enable the verification of the creators of the videos), are obligations of content generation service providers.

12.7 SUGGESTIONS AND THE ROLE OF THE INDIAN COMPUTER EMERGENCY TEAM (CERT-IN)

After having a look toward the threat that deepfakes pose to the security of India and analyzing the current legislation in this domain across the globe, and by scrutinizing the current Indian legislations that indirectly fight with deepfakes, we are of the opinion that Indian law lags behind in tackling the harms associated with deepfakes. Since law is a social science, it advances and changes along with societal changes. Law is necessary to address the difficulties brought about by new social developments. The concept and application of law must evolve together with the society if it is to remain relevant [51]. Hence, we are going to provide some suggestions to lawmakers and the government of India on how to respond to the harms associated with deepfakes to the security of India:

1. A specific legislation related to the harms associated with deepfakes and AI shall be introduced and passed in the Parliament, just like other countries across the globe did, and the term "deepfake" shall be defined in that legislation.
2. That legislation should not hamper innovations and constructive development in the domain of AI.

3. That law should address all the concerns related to the security of India, as discussed till now, and provide for robust mechanisms to deal with these harms.
4. That legislation shall make compulsory for intermediaries to remove deepfake content from their platforms, as soon as possible, and failing to comply with these provisions shall attract penalties.
5. The government will have to tie up with tech giants, like Google and Microsoft, in order to invest and promote research and development in detecting deepfakes and shall promote the use of existing deepfake detecting tools. Certain tools that can detect deepfake are as follows:
   a. The Microsoft Video Authenticator tool can spot grayscale details and blending borders that the human eye cannot see.
   b. A generative AI model's fingerprints are picked up by Facebook reverse engineering [52].

Moreover, the Indian Computer Emergency Response Team (CERT-In) has to play a major role in detecting deepfakes. CERT-In will also have to organize seminars and programs to educate the people of India about the threat that deepfakes pose, what they can do when they become victim to cybercrime, and so on. It has to invest heavily on forensics detection techniques and must partner with other tech giants and stakeholders in this area.

12.8 CONCLUSION

Deepfakes are still a threat to people and businesses, with the potential to have significant effects on entire countries, governments, corporations, and societies. Professionals from various fields whose research interests coincide with deepfakes generally concur that technology is developing quickly, the expensive cost of creating high-quality deepfake content is going down, and the converse is true for deepfake detection technology. The expense of detecting deepfakes is very costly, and the technology for doing so is not developing at such a rapid pace. We therefore anticipate an evolving threat landscape in which attacks will grow more frequent and effective, and in which government agencies, business sectors, and civil society will need to coordinate their efforts in order to resist and mitigate these risks.

At last, according to our opinion, technology and law should work together, and the time has come to promote and work on the blending of technology and law, that is, techno-legal, to fulfill the principles of our holy Constitution.

REFERENCES

1. Besson, S. (2011, April). Sovereignty. *Oxford Public International Law*. https://opil.ouplaw.com/display/10.1093/law:epil/9780199231690/law-9780199231690-e1472.
2. Schaap Arie, J. (2009). Cyber Warfare Operations: Development and Use Under International Law. *Air Force Law Review*, 64(68), 121–173. Who defines 'cyberspace' as a "domain characterized by the use of [computers and other electronic devices] to store, modify, and exchange data via networked systems and associated physical infrastructures".

3. UNGA. (2013). Group of Governmental Experts on Developments in the Field of Information and Telecommunications in the Context of International Security, 24 June 2013, UN Doc A/68/98, para 20; UNGA. (2015). Report of the Group of Government Experts on Developments in the Field of Information and Telecommunications in the Context of International Security, A/70/174, 22 July 2015, para 27.
4. U.S. Department of Defense, Cyberspace Policy Report. (2011, November). *A Report to Congress Pursuant to the National Defense Authorization Act for Fiscal Year 2011, Section 934*. www.defense.gov/home/features/2011/0411_cyberstrategy/docs/NDAA%20Section%20934%20Report_For%20webpage.pdf.
5. Kesavananda Bharti v. State of Kerala, AIR 1973 SC 1461.
6. Cover, R. (2022). Deepfake Culture: The Emergence of Audio-Video Deception as an Object of Social Anxiety and Regulation. *Journal of Media & Cultural Studies*, 4–12.
7. Gaur, L. (2023). *DeepFakes Creation, Detection, and Impact*. Taylor & Francis.
8. Fegan, K. (2018, April 18). A Video that Appeared to Show Obama Calling Trump a "DIPSH-T" is a Warning about a Disturbing New Trend Called 'Deepfakes'. *Business Insider India*. www.businessinsider.in/tech/a-video-that-appeared-to-show-obama-calling-trump-a-dipsh-t-is-a-warning-about-a-disturbing-new-trend-called-deepfakes/articleshow/63807263.cms.
9. Parkin, S. (2019, June 22). Politicians Fear This Like Fire. *The Guardian*. www.theguardian.com/technology/ng-interactive/2019/jun/22/the-rise-of-the-deepfake-and-the-threat-to-democracy.
10. Towers, C. (2019, May 31). Mona Lisa and Nancy Pelosi: The Implications of Deepfakes. *Forbes*. www.forbes.com/sites/charlestowersclark/2019/05/31/mona-lisa-and-nancy-pelosi-the-implications-of-deepfakes/?sh=5e46695e4357.
11. Hao, K. (2020, September 20). Deepfake Putin is Here to Warn Americans about Their Self-Inflicted Doom. *MIT Technology Review*. www.technologyreview.com/2020/09/29/1009098/ai-deepfake-putin-kim-jong-un-us-election/.
12. Allyn, B. (2022, March 16). Deepfake Video of Zelenskyy could be 'Tip of the Iceberg' in Info War, Experts Warn. *NPR*. www.npr.org/2022/03/16/1087062648/deepfake-video-zelenskyy-experts-war-manipulation-ukraine-russia.
13. Blakkarly, J. (2019, June 17). A Gay Sex Tape is Threatening to End the Political Careers of Two Men in Malaysia. *SBS News*. www.sbs.com.au/news/a-gay-sex-tape-is-threatening-to-end-the-political-careers-of-two-men-in-malaysia.
14. Henry, M. (2021, June 24). The Tom Cruise Deepfake that Set Off 'Terror' in the Heart of Washington DC. *Australian Broadcasting Corporation News*. www.abc.net.au/news/2021-06-24/tom-cruise-deepfake-chris-ume-security-washington-dc/100234772.
15. Cuthbertson, A. (2022, November 30). Mark Zuckerberg Deepfake Calls Out Congress for Inaction Over Monopolies. *The Independent*. www.independent.co.uk/tech/mark-zuckerberg-deepfake-ai-meta-b2236388.html.
16. Burton, B. (2020, June 20). MIT Releases Deepfake Video of 'Nixon' Announcing NASA Apollo 11 Disaster. *CNET*. www.cnet.com/science/mit-releases-deepfake-video-of-nixon-announcing-nasa-apollo-11-disaster/.
17. Rahim, Z. (2020, February 25). 'Deepfake' Queen Delivers Alternative Christmas Speech, in Warning about Misinformation. *CNN*. https://edition.cnn.com/2020/12/25/uk/deepfake-queen-speech-christmas-intl-gbr/index.html.
18. Jee, C. (2020, February 19). An Indian Politician is Using Deepfake Technology to Win New Voters. *MIT Technology Review*. www.technologyreview.com/2020/02/19/868173/an-indian-politician-is-using-deepfakes-to-try-and-win-voters/.
19. Ajmal, A. (2020, November 10). Adult Deepfakes of Indian Film Stars Thrive Online. *The Times of India*. https://timesofindia.indiatimes.com/india/adult-deepfakes-of-indian-film-stars-thrive online/articleshow/79140509.cms.

20. Anil Kapoor v. Simply Life India & Ors. CS(COMM) 652/2023 and I.A. 18237/2023–18243/2023.
21. Delfino. R. (2022). Deepfakes on Trial: A Call to Expand the Trial Judge's Gatekeeping Role to Protect Legal Proceedings from Technological Fakery. *Loyola Law School*, 1–6.
22. Deepfakes Accountability Act. *Congress.gov*. www.congress.gov/bill/117thcongress/house-bill/2395/text.
23. Metz, C. (2019, November 24). Internet Companies Prepare to Fight the 'Deepfake' Future. *The New York Times*. www.nytimes.com/2019/11/24/technology/tech-companies-deepfakes.html.
24. Pfefferkorn, R. (2020). "Deepfakes" in the Courtroom. *Public Interest Law Journal*, 245–250.
25. Intel. (2022, November 14). Intel Introduces Real-Time Deepfake Detector. *Intel Newsroom*. www.intel.com/content/www/us/en/newsroom/news/intel-introduces-real-time-deepfake-detector.html#gs.notfp6.
26. Tehseen S. Poonawala v. Union of India, (2018) 9 SCC 501.
27. Sri Adi Visheshwara of Kashi Vishwanath Temple v. State of U.P., (1997) 4 SCC 606.
28. Chatterjee, A. (2023, February 06). Explained | What are Voice Deepfakes and How are They Used? *The Hindu*. www.thehindu.com/sci-tech/technology/explained-voice-deep-fakes-how-are-they-are-used/article66476423.ece.
29. Byman, D., Gao, C., & Meserole, C. (2023). *Deepfake and International Conflict*. Brookings Institution, 8.
30. Jee, C. (2020). An Indian Politician is Using Deepfake Technology to Win New Voters. *MIT Technology Review*. www.technologyreview.com/2020/02/19/868173/an-indian-politician-is-using-deepfakes-to-try-and-win-voters/.
31. The Quint. (2020, February 19). *Delhi Polls: Manoj Tiwari Used Deepfake to Reach Larger Voter Base*. www.thequint.com/news/webqoof/delhi-elections-bjp-manoj-tiwari-used-deepfake-to-reach-larger-voter-base.
32. Byman, D., Gao, C., & Meserole, C. (2023). *Deepfake and International Conflict*. Brookings Institution, 8.
33. Carothers, T., & O'Donohue, A. Political Polarization in South and Southeast Asia. *Carnegie Endowment for International Peace*. https://carnegieendowment.org/2020/08/18/mounting-majoritarianism-and-political-polarization-in-india-pub-82434.
34. VMWare, Global Incident Response Threat Report. (2022). www.vmware.com/content/dam/learn/en/amer/fy23/pdf/1553238_Global_Incident_Response_Threat_Report_Weathering_The_Storm.pdf, last accessed—2023/11/05.
35. Umawing, J. The Face of Tomorrow's Cybercrimes: Deepfake Ransomware Explained. *Malwarebytes*. www.malwarebytes.com/blog/news/2020/06/the-face-of-tomorrows-cybercrime-deepfake-ransomwareexplained.
36. Benscoter, J. (2021, March 17). Woman Created 'Deepfake' Videos to Force Rivals OffDaughter'sCheerleadingSquad.*PennLive*.www.pennlive.com/news/2021/03/pa-woman-created-deepfake-videos-to-force-rivals-off-daughters-cheerleading-squad-police.html.
37. Tessian. (2021, December 26). What are Deepfakes? Are They a Security Threat? *Tessian*. www.tessian.com/blog/whatare-deepfakes/.
38. Byman, D., Gao, C., & Meserole, C. (2023). *Deepfake and International Conflict*. Brookings Institution, 8.
39. Times Fact Check. (2019, September 19). Fake Alert: No, This Video isn't of the Latest Skirmish between Indian and Chinese Soldiers. *Bangalore Mirror*. https://bangaloremirror.indiatimes.com/news/india/fake-alert-no-this-video-isnt-of-the-latest-skirmish-between-indian-and-chinese-soldiers/articleshow/71205096.cms.
40. Kar, M., & Sahoo, S. (2020). Deepfakes and Its Iniquities: Regulating the Dark Side of AI. *NLUO Student Law Journal*, 5, 41.
41. The National Security Act, 1980 (Act 65 of 1980).

42. Subhash Bhandari v. District Magistrate, (1987) 4 SCC 658.

43. Dahiya, R. Disinformation as Threats to National Security and Possible Solutions. *Fake News*. https://ijlpp.com/fake-news-disinformation-as-threats-to-national-security-and-possible-solutions/#_edn11.

44. Pandey, S., & Jadhav, G. (2023, March 17). Emerging Technologies and Law: Legal Status of Tackling Crimes Relating to Deepfakes in India. *SCC Online Blog*. www.scconline.com/blog/post/2023/03/17/emerging-technologies-and-law-legal-status-of-tackling-crimes-relating-to-deepfakes-in-india/#:~:text=Under%20the%20Information%20Technology%20(Intermediary,these%20crimes%20can%20be%20prosecuted.

45. Deepfakes Accountability Act. *Congress.gov*. www.congress.gov/bill/117thcongress/house-bill/2395/text.

46. Strout, N. (2019, November 13). How the Pentagon is Tackling Deepfakes as a National Security Problem. *C4ISRNET*. www.c4isrnet.com/information-warfare/2019/08/29/how-the-pentagon-is-tackling-deepfakes-as-a-national-security-problem/.

47. Briscoe, S. (2021, January 12). U.S. Laws Address Deepfakes. *Asia Online*. www.asisonline.org/security-management-magazine/latest-news/today-in-security/2021/january/U-S-Laws-Address-Deepfakes/.

48. Aseri, A. (2020). Overhauling Publicity Rights vis-à-vis Deepfakes—Truth Disrupted. *Journal of Intellectual Property Law*, 5, 15.

49. Patil, H. (2020, June 20). Deepfakes: A Crime in Cyberspace. *LinkedIn*, www.linkedin.com/pulse/deepfakes-crime-cyberspace-harshvardhan-patil-.

50. Hemrajani, A. (2023, March 08). China's New Legislation on Deepfakes: Should the Rest of the Asia Follow Suit? *The Diplomat*. https://thediplomat.com/2023/03/chinas-new-legislation-on-deepfakes-should-the-rest-of-asia-follow-suit/.

51. Mahajan, V.D. (2022). Jurisprudence & Legal Theory 23 (Eastern Book Company, 34, Lalbagh, Lucknow-226 EBC Publishing (P) Ltd., Lucknow 6th edn.).

52. Barker, P. (2022, September 09). CERT Data Scientists Probe Intricacies of Deepfakes. *IT World Canada*. www.itworldcanada.com/article/cert-data-scientists-probe-intricacies-of-deepfakes/501851.

Index

Printed in the United States
by Baker & Taylor Publisher Services